Weather: A Concise Introduction

From a world-renowned team at the Department of Atmospheric Sciences at the University of Washington, Seattle, *Weather: A Concise Introduction* is an accessible and beautifully illustrated text covering the foundations of meteorology in a concise, clear, and engaging manner. Designed to provide students with a strong foundation in the physical, dynamical, and chemical processes taking place in the atmosphere, this introductory textbook will appeal to students with a wide range of mathematical and scientific backgrounds.

This textbook provides a practical approach to the study of meteorology. It features: a single case study of a midlatitude cyclone which is referred to throughout the whole book to illustrate the basic principles driving atmospheric dynamics and phenomena; boxes on more advanced topics; appendices for additional coverage; chapter summaries listing the "take-home" points discussed; and color figures and charts clearly illustrating the fundamental concepts. Key terms are evident throughout, and a glossary explains the terms that students will need to understand and become familiar with.

Gregory J. Hakim has undergraduate degrees in Mathematics and Atmospheric Science and a PhD in Atmospheric Science from the University at Albany, State University of New York. He joined the Department of Atmospheric Sciences at the University of Washington in 1999, where he served as Department Chair from 2012 to 2017 and is currently a Professor. He is also a leading scientist in the areas of weather analysis, predictability, and dynamics, and his research interests include weather and climate prediction, hurricanes, past climates, and polar circulation patterns.

He has served on the advisory panel for the Directorate of Geosciences at the National Science Foundation, as Chair of the advisory panel for the Mesoscale and Microscale Meteorology Laboratory at the National Center for Atmospheric Research (NCAR), as a member of the NCAR Advisory Panel, as a member of the NCAR Strategic Planning Council, and as Chair of the University Corporation for Atmospheric Research's President's Advisory Committee on University Relations.

Jérôme Patoux earned a Master in Environmental Engineering from the University of Texas at Austin and a PhD in Atmospheric Science from the University of Washington. He has been funded by the National Science Foundation (NSF), the National Aeronautics and Space Administration (NASA), the Office of Naval Research (ONR), and the National Oceanic and Atmospheric Administration (NOAA). He has taught undergraduate introductory meteorology for many years, and has been funded by the NSF to develop weather and climate curriculum. He is a former faculty member from the Department of Atmospheric Sciences at the University of Washington, and currently teaches meteorology at the University of Nantes in France.

Weather

A Concise Introduction

GREGORY HAKIM
University of Washington

JÉRÔME PATOUX
University of Washington

CAMBRIDGE
UNIVERSITY PRESS

CAMBRIDGE
UNIVERSITY PRESS

University Printing House, Cambridge CB2 8BS, United Kingdom

One Liberty Plaza, 20th Floor, New York, NY 10006, USA

477 Williamstown Road, Port Melbourne, VIC 3207, Australia

314–321, 3rd Floor, Plot 3, Splendor Forum, Jasola District Centre, New Delhi – 110025, India

79 Anson Road, #06–04/06, Singapore 079906

Cambridge University Press is part of the University of Cambridge.

It furthers the University's mission by disseminating knowledge in the pursuit of education, learning, and research at the highest international levels of excellence.

www.cambridge.org
Information on this title: www.cambridge.org/9781108417167
DOI: 10.1017/9781108264983

© Gregory Hakim and Jérôme Patoux 2018

First published 2018
Reprinted 2019

Printed in Singapore by Markono Print Media Pte Ltd

A catalogue record for this publication is available from the British Library.

ISBN 978-1-108-41716-7 Hardback
ISBN 978-1-108-40465-5 Paperback

Contents

Preface

Having taught introductory classes on weather many times, we came to see the need for a textbook on the subject that covers the foundations of meteorology in a concise, clear, and engaging manner. We set out to create an informative, cost-effective text that meets the needs of students who may not have any background in mathematics and science. The result – *Weather: A Concise Introduction* – is an introductory meteorology textbook designed from scratch to provide students with a strong foundation in the physical, dynamical, and chemical processes taking place in the atmosphere.

This textbook is unique in that it:

▶ provides a concise and practical approach to understanding the atmosphere;

▶ introduces the basic physical laws early on and then ties them together with a single case study spanning the book;

▶ presents weather analysis tools early in the book to allow instructors to engage in discussions of current weather in tandem with the basic concepts, thus attracting and retaining student interest; and

▶ facilitates students' learning and understanding of the fundamental aspects of weather analysis and forecasting, as well as practical skills, through a careful description of the forecasting process. Modern methods, such as ensemble forecasting, are central to the approach.

Features

Case Study: February 2014 Cyclone

The main concepts of the book are illustrated in Chapters 2–13 by a single case study: a midlatitude cyclone that swept through the eastern half of the USA between February 19 and 22, 2014. This rich case study serves as a common thread throughout the book, allowing students to study it from multiple perspectives. Viewing the storm in the context of different topics provides a familiar setting for mastering new subjects and for developing an holistic understanding of midlatitude cyclones.

Boxes on More Advanced Topics

Instructors have the option of including more advanced coverage through use of boxes that provide insights on various topics. For example, in Chapter 1, Weather Variables, boxes include an in-depth description of the four laws of physics that are central to the study of the atmosphere. The book contains 25 boxes, affording instructors the opportunity to tailor the level of the material that they present to students in their course.

Appendixes for Additional Coverage

Appendixes at the ends of Chapters 2, 3, 6, 7, and 10 include additional material on important cloud signatures found in satellite imagery, the concept of dynamic equilibrium, the cloud classification, some optical phenomena, southern hemisphere midlatitude cyclones, and the Bergen School of meteorology.

Summary

A summary of key points has been included at the end of each chapter so that students can, at a glance, confirm that they have understood the significant takeaway facts and ideas.

Figures, Charts, and Maps

Figures have been designed to convey the key concepts in a simple and self-explanatory way, keeping in mind that clean representations of information are more helpful to students than complex drawings. Graphs and maps have been created with real data as much as possible, obtained from NOAA, NASA,

ECMWF, and similar research-quality sources referenced in the text.

Key Terms and Glossary

The main text contains terms (in bold) that students need to understand and become familiar with. Many of these terms are listed in the Glossary at the back of the book. The Glossary allows the reader to look up terms easily whenever needed and can also be used to review important topics and key facts.

SI Units

We have consistently used SI units throughout the book, while providing alternative units whenever possible or relevant.

Organization

The first two chapters provide a general overview of key variables and weather maps used by meteorologists, which facilitates daily weather map discussions early in the course. We have found that motivating lecture topics with real-time examples using weather map discussions is a very effective way to engage students in the lecture material, and it allows instructors to introduce aspects of weather forecasting at their discretion well in advance of discussing the material more completely in Chapter 13. As a result, students are more invested in adding to their knowledge, which builds systematically toward understanding and predicting weather systems.

Chapters 3–8 provide foundational material on the composition and structure of the atmosphere, along with the application of the laws of classical physics to emphasize and explain the role of energy, water, and wind in weather systems.

Chapters 9–12 apply the foundational material to understanding the general circulation of the atmosphere (Chapter 9), midlatitude cyclones and fronts (Chapter 10), thunderstorms (Chapter 11), and tropical cyclones (Chapter 12).

Chapters 13–15 build further on the first twelve chapters by applying the concepts developed to explain processes that affect how weather forecasts are made (Chapter 13), air pollution (Chapter 14), and climate change (Chapter 15).

Instructor Resources

A companion website at www.cambridge.org/weather contains PowerPoint slides of the figures in the text as well as a testbank of questions.

Acknowledgments

We thank: NOAA, NASA, and ECMWF for providing access to data and images; Reto Knutti, Jan Sedlacek, and Urs Beyerle for providing access to IPCC data; Rick Kohrs from the University of Wisconsin-Madison for providing global composite satellite imagery; and Paul Sirvatka from the College of DuPage for providing radar imagery.

We also thank Ángel Adames, Becky Alexander, Ileana Blade, Peter Blossey, Michael Diamond, Ralph Foster, Dargan Frierson, Qiang Fu, Dennis Hartmann, Lynn McMurdie, Paul Markowski, Cliff Mass, Max Menchaca, Yumin Moon, Scott Powell, Virginia Rux, David Schultz, Justin Sharp, Brian Smoliak, Mike Warner, Steve Warren, Rachel White, Darren Wilton, Matt Wyant, and Qi Zhong, as well as 13 anonymous reviewers, for their help in the preparation of this book.

This project would not have come to life without the support, help, influence, and constructive criticism from many fellow professors, teaching assistants, and students. We cannot acknowledge them all here by name, but we thank them nevertheless for the important role they have played in shaping the development of this book.

Introduction

Why should we study our atmosphere? Why should we learn about the causes and mechanisms of our weather? Weather affects our daily life: the clothes we wear (rain coat, shorts, hat, should we take an umbrella or sunglasses...?), the means of transportation we choose (walk, take a bus, ride our bike...?), our activities (ski, sail, water our plants, read a book in a coffee shop...?), and probably more. But beyond our daily concerns, weather affects society at large. Schools close when snow impedes traffic. Visitors to ski resorts might be more impatient for snow, while the ski instructors will be keeping an eye on the possibility of avalanches. Rangers are concerned with fog, thunderstorms, and flash floods. Fire patrols look for weather patterns that are conducive to forest fires (dryness, wind). Electricity providers are concerned by wind storms that can damage the infrastructure of the electrical grid and, on larger timescales, also need to plan how weather will affect upcoming energy needs (minimum temperatures impact heating, while maximum temperatures impact air-conditioning). Weather averages, such as prevailing winds, the typical temperature range, and mean precipitation determine how we build our homes and what locations are sensitive to extreme events, such as droughts, floods, hurricanes, tornadoes, etc. On longer timescales, we can ask how humans are changing the atmosphere, and what those changes imply for the weather and climate of the future.

To start answering those questions, we need to understand how the atmosphere works. We need to identify the basic processes that drive the atmosphere, and the laws that govern atmospheric processes. By doing so, we will be able to explain the weather phenomena we experience around the year and throughout the world. Furthermore, we will also be able to apply these laws to the current state of the atmosphere, and *predict* how it will evolve in the future.

© Caroline Planque

There is a lot of value in becoming a knowledgeable observer of the atmosphere. After reading this book, you will look at the sky differently, you will gain an understanding of weather and climate that will make you more attentive to the world around you. You will have a basic understanding of weather phenomena, of cyclones, thunderstorms, and hurricanes, and you will understand the basic aspects of weather forecasting. You will see beyond the weather forecast you get on your phone, radio, TV, or the internet, and you will be able to make your own forecast in many situations.

Weather and Climate

Before we continue, let us clarify an important distinction between weather and climate. **Weather** is the *condition of the atmosphere at a particular time and location*. Weather varies on timescales of minutes to days. **Climate**, by contrast, is an *average of the weather*. It varies on timescales of decades to centuries and beyond. In this textbook, we will be mostly concerned with weather – even though many of the concepts have direct application to climate.

Getting Started

Our exploration of weather will start with a quick overview of important weather elements that we can observe or measure, and analyze. The choice of variables to observe is influenced by the laws of physics that govern the atmosphere. As we will see shortly, the atmosphere is made of *matter* (air and water etc.), it contains *energy* (heat), and it is in *motion* (wind, convection). Our understanding of weather is based on the fundamental notion that matter, energy, and motion obey *conservation laws*. To apply these conservation laws to the atmosphere requires observations of temperature (conservation of energy), pressure (conservation of mass), wind (conservation of momentum), along with humidity, precipitation, and clouds. One step at a time, and one building block over another, we will then investigate the physical processes that underlie the atmosphere at work. Finally, we will articulate these processes together to build a picture of weather systems such as mid-latitude cyclones, thunderstorms, and hurricanes. In doing so, we will follow the precepts of René Descartes, who advocated, as early as 1637, that every difficult problem should be divided into small parts, and that one should always proceed from the more simple to the more complex. This cornerstone of the scientific method, still in favor today, will be an important aspect of our exploration as we elaborate a thorough understanding of the atmosphere from its most fundamental constituents at the molecular scale to its most complex inner workings as a system for moving heat at the global scale.

CHAPTER 1

Weather Variables

Where should we start with our study of the atmosphere? How should we first approach the weather? Like many scientists, meteorologists first make observations. Then they raise questions, and try to answer them. In this first chapter, we will quickly describe four of the elements, also called variables, of weather that meteorologists regularly observe, measure, and chart on weather maps, before we return to each of them for a more thorough exploration in subsequent chapters. Three of these elements are fairly intuitive: when concerned with the weather, we like to know how warm or cold it will be (temperature), whether it will be windy or not (wind), and whether it will rain or not (precipitation). The fourth variable, atmospheric pressure, is less intuitive, but it may be the most important to a meteorologist, as we will soon discover.

Weather results from atmospheric changes. These changes obey certain rules, dictated by the laws of physics. In meteorology, three laws are of particular importance: the law of conservation of energy, the law of conservation of mass, and the law of conservation of momentum. Each describes a particular aspect of the atmosphere, and each requires that we measure certain **variables** of the atmosphere. The object of this first chapter will be to provide an overview of these variables, a starting point for our exploration of atmospheric changes. We will then return to each variable in subsequent chapters for a more thorough description and analysis.

1.1 Temperature

Of primary interest to us is the law of conservation of energy (see Box 1.1). It states that energy is never created or destroyed, but only transferred between

Box 1.1.   The Law of Conservation of Energy

The law of conservation of energy states that the total energy of a system remains constant, if we account for the gains and losses of energy from and to the outside. Energy can be *transferred* between different parts of the system, or *transformed* into different types of energy inside the system (e.g., from potential energy to kinetic energy, or chemical energy, or radiative energy), but energy cannot be created (out of nothing) or destroyed.

If we think of the atmosphere as a system, we can apply the law of conservation of energy to describe how energy, and in particular heat, is transferred or transformed in the atmosphere. And since heat transfers are related to temperature differences, we need to measure temperature and map these differences. That is one reason why temperature measurements are an integral part of weather observations.

We will return to the conservation of energy in Chapter 6, in the form of the first law of thermodynamics.

locations or transformed between different types of energy. In the atmosphere one form of energy is **heat**, and weather is largely the result of heat contrasts and heat transfers. Therefore we need to design ways of describing the amount and fluxes of heat throughout the atmosphere, which is accomplished by measuring **temperature**.

1.1.1 Heat and Temperature

We all have an intuitive feel for temperature, for whether things are hot or cold. But that feeling is very subjective – warm water feels cold after stepping out of a sauna. Science requires objectivity, which is sought by measuring and quantifying variables and processes. But how do we quantify that feeling of warm and cold? What is temperature, really? First, we need to return to the fundamental definition of **heat** as it is transferred to our body and the environment by contact and interactions at the molecular level.

We can think of the atmosphere as a mixture of gases made up of molecules in motion (Figure 1.1). And we can think of heat as the energy associated with this molecular motion. In warm air, molecules are moving more rapidly than in cold air, and therefore have more energy of motion. We call this energy of motion, **kinetic energy**. Temperature is an indirect

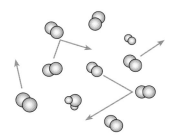

Figure 1.1.   Random molecular motion in a volume of air.

measurement of the *average* kinetic energy of the molecules constituting the air. Here it is useful to think of the air around us as made of individual volumes of air of uniform characteristics, called *air parcels*. [Although the size of such volumes of air is somewhat arbitrary, the concept of an air parcel will be useful later to explain various processes at work in the atmosphere.] While individual molecules in the parcel might all have different speeds, their average state is indicative of a higher or lower energy level, hence a higher or lower temperature of the air parcel.

When fast-moving air molecules (at a higher temperature) are in contact with slow-moving air molecules (at a lower temperature), they impart some of their energy to the slow molecules, by contact, through collisions. In doing so, they *lose* energy, and therefore cool down, while the slower molecules *gain* energy, and therefore warm up. Kinetic energy, and therefore heat, has been transferred from the faster air molecules to the slower molecules, from the warmer to the colder air. We will see in Chapter 4 that this is the basic mechanism for transferring energy at the molecular level through **conduction**, but it is not a primary heat transfer mechanism for most of the atmosphere.

1.1.2 Thermometers

How do we measure temperature? We make indirect measurements of average kinetic energy using **thermometers**. Liquid-in-glass thermometers, for example, work on the basis that substances expand or contract when temperature increases or decreases. A glass tube is filled with a liquid such as alcohol or mercury. When placed in warmer air, the energy of motion of the air molecules is transferred to the liquid in the reservoir by conduction through the glass. The molecules constituting the liquid now have more energy than before. Being more active, they push each other apart, which makes the liquid expand. In particular, it makes the liquid rise in the tube. Exposed to the same temperature, the liquid will always rise to the same level. Thus, if we attach a scale and units to the glass tube, and calibrate the instrument against known temperatures (such as the freezing point and the boiling point of water), we obtain an instrument that can measure any temperature (Figure 1.2).

Many other types of thermometer exist, and all make use of a property of matter to determine temperature indirectly. A bimetallic strip, for example, is made of two thin pieces of different types of metal attached to each other. Because the metals expand and contract at different rates, the combined strip bends when the temperature changes. If the tip of the metal strip is made to bend toward a temperature scale, the device can be calibrated and turned into an instrument.

Some electronic thermometers use a material whose electrical resistance depends on temperature (a thermistor). Others, called radiometers, measure the radiation emitted by bodies. Because the emitted radiation is a function of the temperature of these bodies, we can once again indirectly deduce their temperature, as we will describe in Chapter 4.

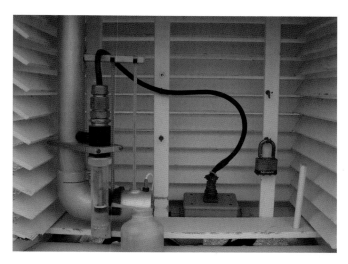

Figure 1.2. Thermometer in a standard weather shelter.

1.1.3 Temperature Measurements

As we will discuss at the end of this chapter, temperature measurements, like other weather measurements, are standardized, so that they can be compared and mapped. Weather stations always measure temperature in the shade, to avoid contamination by sunlight – recall that we are interested in measuring the kinetic energy of the air molecules surrounding the thermometer, and not the radiative energy contained in sunlight that might be absorbed by the thermometer. For this reason, temperatures are measured in a shelter, which is elevated at 2 meters (6½ feet) height above a vegetated surface, to avoid contamination by surface effects. Indeed, the air temperature can change rapidly near the ground, even within 2 meters (see Chapter 4 and Figure 4.14 in particular), and it is important that we measure the temperature of the *air*, not that of the surface. The shelter is painted white, and ventilated, to limit the absorption of sunlight and the concentration of heat inside the box, which could otherwise produce an artificially high temperature.

1.1.4 Temperature Scales

You are probably familiar with the Fahrenheit temperature scale, named after Gabriel Fahrenheit, a German scientist who constructed the first mercury thermometer in the eighteenth century and calibrated it against three fixed points: 0, as defined by a mixture of ice, water, and sea salt; 32 in water and ice; and 96 "in the mouth or armpit of a healthy man." Degrees Fahrenheit are commonly used in the United States. Scientists, and most other countries of the world, prefer to use the Celsius (or Centigrade) temperature scale, named after Anders Celsius, a Swedish astronomer who also lived in the eighteenth century and proposed the temperature of melting ice and the temperature of boiling water as fixed points (0 and 100, respectively). To convert from one scale to another, use the following formulas:

$$°C = 5/9 \times (°F - 32°)$$
$$°F = 9/5 \times (°C) + 32°$$

(Note that, in the first formula, 32 is subtracted *before* multiplying by 5/9, while, in the second formula, 32 is added *after* multiplying by 9/5 – a common source of mistake.)

Scientists also use the Kelvin temperature scale, named after Lord Kelvin, a nineteenth-century scientist whose original name was William Thomson. The Kelvin scale is merely an offset version of the Celsius scale, translated in such a way that temperature measurements are always positive numbers.

$$K = °C + 273.15°$$
$$°C = K - 273.15$$

The smallest possible temperature, zero kelvin (0 K), is called "absolute zero," and corresponds to the theoretical state in which all molecular motion stops, in which case molecules have zero kinetic energy, and therefore zero temperature.

(Note that the symbol for kelvin (K) does not have the degree symbol – the little circle – in contrast with °F and °C.)

Here, in keeping with the International System of Units, we use degrees Celsius on weather maps and temperature profiles. If you are more accustomed to degrees Fahrenheit, however, it will be useful to be able to convert from one scale to another, using the formulas above, or Table 1.1 for quick reference. If you hear or read temperatures in degrees Celsius and

Table 1.1. *Common temperature values in degrees Celsius and Fahrenheit*

°C	°F	°C	°F
−40	−40	**0**	**32**
−35	−31	5	41
−30	−22	10	50
−25	−13	15	59
−20	−4	20	68
−15	5	25	77
−10	14	30	86
−5	23	35	95
0	**32**	40	104

want to convert them quickly to degrees Fahrenheit, you can take advantage of the fact that 9/5 is approximately equal to 2, and 32 is approximately equal to 30. Thus, you can use the following approximation:

$$°F \approx 2 \times °C + 30$$

This is a very quick calculation that you can easily do in your head: multiply by 2 and add 30. For example, 20 °C multiplied by 2 is equal to 40. Add 30 to obtain 70 °F, which is close enough to the exact conversion, 68 °F.

1.1.5 Radiosonde Profiles

As we will see later in the book, a large part of our weather is dictated by what takes place aloft, as weather systems often extend up into the atmosphere and surface phenomena are often driven by upper air currents. Therefore it is useful to obtain information about the vertical structure of the atmosphere (Figure 1.3). We do so by launching helium-filled balloons carrying instruments that record weather variables up to 35 km altitude as the balloons ascend – helium is used because it is a very light gas (Figure 1.4). These balloons are called **radiosondes**, since the measurements are radioed back to a receiving station at the surface.

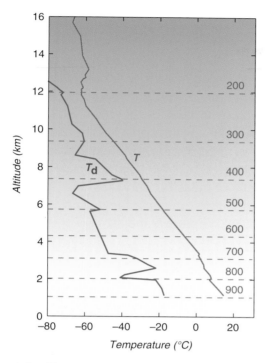

Figure 1.3. Temperature profile obtained by radiosonde at Amarillo, Texas, on February 18, 2014, at 00:00 UTC. The red curve indicates temperature (T) and the green curve indicates the dew point temperature (T_d). Pressure levels (given in hectopascals, hPa) are shown in blue and will be described in the text.

Figure 1.4. Radiosonde being launched in Hawaii.

Figure 1.3 shows the lower section of a temperature profile, in red, obtained by launching a radiosonde from Amarillo, Texas, on February 18, 2014, at 00:00 UTC. [Except where specified, we use Universal Time Coordinates (UTC) in the rest of the book, i.e., time referenced at the meridian of Greenwich, England.] The green curve indicates the dew point temperature, an important measure of the amount of water vapor in the air, which we will discuss in Chapter 5. Figure 1.3 also gives us an opportunity to describe a particular type of graph that we will encounter many times in this book. In such a graph, altitude is displayed on the y-axis (i.e., the vertical axis), while the variable of interest (temperature, pressure, wind speed) is displayed on the x-axis. Thus, the red curve in Figure 1.3 tells us that the temperature decreases from about 18 °C at 1 km altitude to −60 °C at 12 km altitude, which is in fact fairly typical. For reference, pressure levels are indicated by dashed blue lines, labeled in hectopascals (see Section 1.2.5). We can see that the temperature decreases up to about 200 hPa. We will soon learn that this level of the atmosphere (which changes with time and location) is called the *tropopause* and defines the upper limit of the *troposphere* – the layer of the atmosphere where essentially all weather takes place.

1.2 Pressure

Atmospheric changes, and therefore weather, involve the movement of air, and the redistribution of air mass in the atmosphere. Such redistribution obeys the law of conservation of mass, which, in meteorology, is expressed in terms of pressure (see Box 1.2). It is therefore important that we measure pressure to analyze and predict the weather. As we did for temperature, however, it is useful that we first understand the nature of pressure in the atmosphere before we describe instruments to measure it.

Box 1.2. The Law of Conservation of Mass

Much in the same way as energy is conserved, matter, and therefore mass, is conserved in a system. It cannot be created out of nothing and cannot be destroyed: it can only enter or leave the system, or be transformed inside the system (by chemical reaction, for example, or by phase changes, as we will discuss in chapter 5). This is described by the law of conservation of mass.

In meteorology, the distribution of mass in the atmosphere is described by the distribution of density, which is defined as mass per volume. However, because density is difficult to measure, we instead rely on measurements of pressure and temperature to infer density (see Box 1.3). As a result, we can consider pressure measurements as informing us about the distribution of mass in the atmosphere.

Notions of mass conservation and pressure will be important for understanding wind and weather systems in Chapters 8 and 10.

1.2.1 Force and Pressure

As we discussed for temperature, air is a gas made up of molecules in motion, and heat can be equated with the kinetic energy of the molecules. However, we also recognized that these molecules constantly bump into each other. Loosely speaking, the amount of bumping is what we call **pressure**. (We will come to a more exact definition shortly.) If the molecules bump into each other more frequently, or if the collisions themselves are stronger, there is more pressure. Thus, in a closed container, we could increase the pressure by either decreasing the size of the container at fixed temperature, which would bring the molecules closer to each other and would increase the number of the collisions, or increasing the temperature by heating the air, which would provide more kinetic energy to the gas molecules, and would result in more intense collisions. We can see here that pressure, temperature, and the volume of the container are indeed related, as expressed by the ideal gas law (see Box 1.3). However, this is not to say that pressure and temperature are the same thing. Temperature, as a measure of heat, relates to the average kinetic energy of the molecules, independently of the collisions. Conversely, pressure relates to the collisions between molecules, which is not uniquely determined by the speed of the molecules, i.e., it is also a function of the number of molecules in the volume, and therefore the density of the air.

In the atmosphere, things are slightly more complicated, as the air is not enclosed in a container, but we will address that issue shortly. In the meantime, we can think of pressure as the amount of bumping between molecules. And since the molecules are moving in all directions, we can see that pressure is applied in all directions as well.

In practice, we often need to know what happens when air pushes on particular surfaces, either real, like the ground, or imaginary. For example, we sometimes think of the atmosphere as being made of air columns standing next to each other, and we are interested in knowing how much the air in two adjacent columns is pushing against the "wall" in between the columns. At other times, as described earlier, we will think of the atmosphere as being made of air parcels, i.e., volumes of air delimited by an imaginary envelope. Then we will be interested in knowing how much the air inside and outside the parcel pushes on the envelope. In all cases, pressure is applied to a surface, which results in a **force** pushing on that surface. As the area on which pressure is applied increases (i.e., more air molecules bump into that surface), the force increases proportionately. In symbolic notation:

$$F = p \times A$$

where p stands for pressure applied to a surface of area A, and F is the resulting force. Equivalently, pressure can be thought of as the force applied to a surface divided by the area of the surface:

Box 1.3. The Ideal Gas Law

To very good approximation, the atmosphere behaves like an ideal gas, in which molecular collisions result only in a transfer of kinetic energy. As a result, pressure, temperature, and density do not vary independently of each other, but are related by the ideal gas law,

$$pV = nR^*T$$

where p stands for pressure, V denotes volume, n is the number of molecules in volume V, and T is temperature. R^* is a universal constant that applies to any gas.

For the atmosphere, the ideal gas law is more conveniently written:

$$p = \rho RT$$

where ρ is the density (mass per volume) and R is a gas constant specific to the atmosphere.

For example, if we heat a fixed volume and mass of air (i.e., the density remains constant), the kinetic energy of the molecules increases (i.e., the temperature increases), and the molecules bump into each other with more impetus (i.e., the pressure also increases).

$$p = F/A$$

which is, in reality, how pressure is defined in physics. In the above formula, if the force is distributed over a large area, the resulting pressure is small. But if the force is concentrated on a small area, the pressure is high.

A simple application comes to mind. If you are wearing regular shoes, your weight, which provides the force, is distributed over the entire sole of your shoes, which results in a small pressure being applied to the floor. With high heels, however, weight is concentrated on a very small area, which results in a much greater pressure, and could dent a soft floor surface.

1.2.2 Atmospheric Pressure

By analogy, we now consider the weight of the entire atmosphere applied onto Earth's surface at sea level. Air might seem weightless, as philosophers of antiquity, including Aristotle, used to think, but gravity pulls air downward, keeping the atmosphere, like everything else, around Earth. In a column of air extending from Earth's surface all the way up to the top of the atmosphere, the accumulated weight of all the air pushes down on the surface. Recall that weight is a force, so when we divide the weight of the

air column by the area of the base of the atmospheric column, we obtain the pressure of the atmosphere, or **atmospheric pressure**.

Note that, in this explanation, we are thinking about atmospheric pressure as pushing essentially downward, but recall that pressure at a point acts equally in all directions.

1.2.3 Vertical Distribution of Pressure

We do not need to be at sea level for the concept of atmospheric pressure to make sense. We can repeat the same exercise at some altitude above sea level, say, 3000 m. Air pressure at that altitude is determined by how much the air is compressed due to the accumulated weight of the overlying layers of air. (Note that the amount of air below 3000 m is irrelevant in calculating pressure *at* 3000 m.) Since there is necessarily less air above 3000 m than there is above sea level, air pressure will be less than at sea level (Figure 1.5).

To make a familiar connection with this decrease in pressure with altitude, notice the feeling you experience when you swim to the bottom of a pool – you experience an increase in pressure in your ears and nose. The atmosphere is like a swimming pool, and we live at the bottom of the pool – a pool of air. In the same way that a diver will experience higher pressure at depth in the ocean, atmospheric pressure is highest

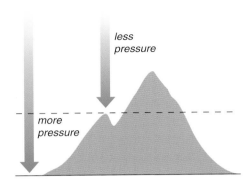

Figure 1.5. Atmospheric pressure is tantamount to the weight of the atmosphere above us.

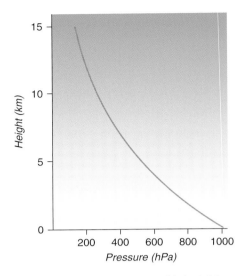

Figure 1.6. Pressure decreases with height.

at ground level, and decreases upward (Figure 1.6 – pressure units are defined in Section 1.2.5).

In summary, it is useful to think of atmospheric pressure as the weight of air above a point, or above a unit area. If we are at sea level, we are experiencing the weight of the total atmospheric column, and therefore higher pressure. If we are higher up in the atmosphere, the column above us contains less air than at sea level, and as a result the pressure is lower.

1.2.4 Barometers

As the weather changes, air moves around, resulting in atmospheric pressure changes at a given location. We monitor these changes with an instrument that measures atmospheric pressure, which we call a **barometer**.

In the seventeenth century, Evangelista Torricelli, an Italian scientist who was greatly influenced by the writings of Galileo, invented the principle of the **mercury barometer**. He filled a glass tube with mercury, inverted it in a cistern, and observed that the level of the liquid in the tube would not completely drop, but would stabilize at a certain height above the cistern (about 760 mm). He concluded that the air pressure applied downward onto the mercury in the cistern was forcing the liquid up into the tube to that level. He correctly speculated that the weight of the column of mercury in the tube was balanced by the weight of the atmosphere pressing down on the mercury outside the tube; i.e., the weight of the mercury column is balanced by atmospheric pressure (Figure 1.7).

Torricelli's design remained the basic principle behind most barometers for more than 300 years. If atmospheric pressure increases, the mercury level rises in the tube. Conversely, when a storm approaches and atmospheric pressure drops, a shorter column of mercury is required to balance the weight of the atmosphere, and the mercury level drops. That is why atmospheric pressure is often reported as the height of the mercury column, namely about 76 cm, 760 mm, or 30 inches Hg ("Hg" being the chemical symbol for mercury).

Note that, in reality, Torricelli first experimented with water. However, about 10 meters of water are required to balance the weight of the atmosphere, i.e., about 33 feet, which made the experiment quite cumbersome! Since mercury is much denser, and thus heavier than water, a much shorter column of mercury is required to achieve the same weight.

If your barometer at home is relatively small and round, you must be wondering how it can possibly contain 30 inches of mercury in a tube. And it does not, of course. You probably own an **aneroid barometer**, which is made of a small empty chamber (evacuated of some air to create a partial vacuum) that can expand and contract. When atmospheric pressure increases, it squeezes the chamber and forces it to contract. When atmospheric pressure decreases, the chamber expands. By connecting the chamber to a moving needle, the expanding chamber can be turned into a precise instrument (Figure 1.8).

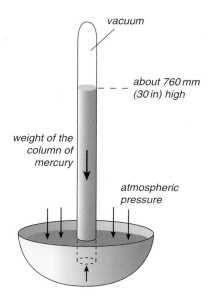

Figure 1.7. Concept of a mercury barometer.

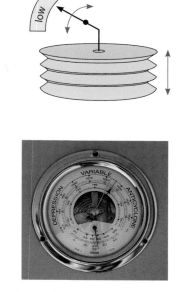

Figure 1.8. Aneroid barometer.

1.2.5 Pressure units

In the International System of Units, pressure is expressed in pascal (Pa), where $1\,Pa = 1\,N/m^2$ (one newton per square meter), or alternatively in kilopascal ($1\,kPa = 1000\,Pa$). Note that a newton is a unit of force and a square meter is a unit of surface area, which is consistent with our earlier definition of pressure as force divided by area. In meteorology, however, since we think of atmospheric pressure as the weight of the atmosphere exerted on a surface, it can be more intuitive to express it in pounds per square inch (psi), or kilograms per square centimeter (kg/cm^2) in the metric system. In that context, we can picture a column of air extending from Earth's surface up to the top of the atmosphere. If the base of the column has a surface area of one square inch, then the weight of that air column is on average 14.6 pounds (at sea level); thus, atmospheric pressure is, on average, 14.6 psi.

Meteorologists, however, go one step further. Since we are interested in the pressure exerted by the atmosphere, we may express pressure... in atmospheres! The average atmospheric pressure would then be close to "one atmosphere." For historical reasons, one atmosphere is called "one bar," after the Greek "baros," meaning weight. Since daily variations in pressure are on the order of a thousandth of an atmosphere (i.e., a thousandth of a bar), meteorologists have historically used

millibars (mb), where 1 bar = 1000 mb. The exact value of a bar has been chosen as 100 kPa for practical purposes, which is equal to 1000 hectopascal (hPa, where hecto- is the prefix for 100), so millibar and hectopascal can be used interchangeably. Putting it all together, the average pressure at sea level is approximately:

$$1013\,hPa = 1013\,mb = 29.92''\,Hg = 14.6\,psi$$

To be consistent with the International System of Units, we will use hectopascals (hPa) in the rest of this book.

1.2.6 Some Useful Numbers

If you observe the variations in atmospheric pressure at a given location over time using a barometer, you will find that the typical range of pressure at sea level is about 980 to 1030 hPa. The pressure will only drop below 980 hPa during intense storms. Similarly, it will only rise above 1030 hPa during exceptionally high pressure events.

If you carry your barometer upward, you will find that pressure drops very quickly with height, as we saw in Figure 1.6. As a rough rule of thumb, pressure near the surface drops about 8 hPa for every 60 m of altitude – this is easily noticeable, and measurable, in elevator trips in tall buildings. Recall the pool analogy: pressure is greatest at the bottom of the pool, or ocean of air, and decreases as we rise in the pool. Since we will be interested in what happens higher up in the atmosphere, it is useful to have some standard levels of pressure in mind and know their approximate altitude.

Figure 1.9 shows some useful numbers to which we will refer often. We saw that the average pressure at sea level is 1013 hPa. We will often look at maps of weather variables at 850 hPa, which is around 1.5 km altitude (about 5000 ft), and near the top of what we call the "boundary layer" (i.e., near the top of the layer where the effects of the daily variations in temperature near Earth's surface are felt). At 500 hPa, about half of the atmosphere is above us and the other half below (in terms of mass and weight), and this happens at about 5.5 km, or 18,000 ft. Aircraft cruising altitude is at about 250 hPa (roughly 11 km, or 34,000 ft), which is also the top of the **troposphere**. Thus, the troposphere, i.e., the first layer of the atmosphere, where most of our weather takes place, contains about 75% of the mass of the atmosphere. Note that 250 hPa also corresponds to the altitude where we find the jet stream, as will be discussed in Chapter 9. Above the troposphere, we enter the stratosphere, where the ozone layer is found, as we will describe in more detail in Chapter 3.

Since water is 1000 times more dense than air, the increase in pressure with depth happens much faster in the ocean than in the atmosphere. Recall that the weight of the atmosphere is balanced by a 10 m column of water in a water barometer. In other words, a 10 m column of water has the same weight as a column of air extending from sea level to the top of the atmosphere. Therefore, we will add the equivalent of another atmosphere of pressure (i.e., one additional bar) if we dive 10 meters down into the ocean – another good number to remember. And for each additional 10 m layer of ocean, we will add another bar of pressure. That is why it is important for divers to pause regularly, especially on their way up to the surface, to let their blood and other internal fluids adjust to the change in external pressure.

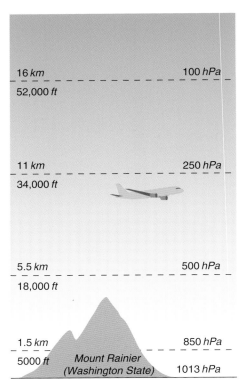

Figure 1.9. Altitude and pressure in the atmosphere: some useful numbers to remember.

1.3 Wind

As mentioned before, weather results from atmospheric motions, and air motion is also constrained by the laws of physics. As for energy and mass, air motion is constrained by a conservation law: the law of conservation of momentum (see Box 1.4). Momentum is the product of mass and velocity. The density of air is estimated indirectly from measurements of pressure and temperature, using the ideal gas law (see Box 1.3). The wind velocity, however, needs to be measured. It has two components, direction and magnitude, which need to be measured separately.

1.3.1 Measuring Wind

Wind direction is measured using a **wind vane**. The vane points toward the direction that the wind is coming from (Figure 1.10), which is why meteorologists report wind direction *as the direction that the wind is*

Box 1.4. The Law of Conservation of Momentum

Weather is a result of the atmosphere in motion to redistribute heat. In meteorology, it is useful to describe air motion in the form of **momentum**, a product of the mass, or density, of the air and its speed, or velocity. Momentum tells us both how much air is in motion, and how fast it moves.

As it turns out, much like energy and mass obey conservation laws, momentum is governed by the law of conservation of momentum, which is described by Newton's laws of motion. In particular, Newton's second law, which relates forces to changes in momentum, describes how the wind evolves. That is why we measure wind as part of our routine weather observations.

Wind and Newton's second law of motion will be the subject of Chapter 8.

coming from. An easterly wind, for example, is a wind blowing from the east. A northerly wind blows from the north, etc. Figure 1.11(a) shows wind in terms of compass direction.

Meteorologists also use degrees from true north to report wind direction (Figure 1.11(b)). Degrees from true north are more convenient for computation and graphing, for example, since they are numerical quantities.

Wind speed is measured with an **anemometer**. The most common type is the cup anemometer (Figure 1.12(a)). As the cups catch the wind, the spindle rotates, and the rate of rotation of the spindle is proportional to the wind speed. It is similar to propeller anemometers, for which the wind turns the blades in proportion to the wind speed (Figure 1.12(b)).

The units for wind speed are miles per hour (mph) in the English system, and kilometers per hour or meters per second in the metric system. (Remember that speed is distance divided by time.) Unfortunately, the most common unit used by meteorologists is knots, for historical reasons: mariners were affected by weather before airplanes were. Sailors used to measure the speed of their ship by letting a knotted rope run into the sea, and counting the number of knots per time as a measure of speed. One knot is one nautical mile per hour, and since one nautical mile is the distance of one minute of latitude on Earth (1.85 km, or 1.15 mi):

$$1\,\text{knot} = 1.85\,\text{km/h} = 1.15\,\text{mph} = 0.5\,\text{m/s}.$$

Mariners sometimes also use the **Beaufort wind scale**, which was first designed to relate wind speed

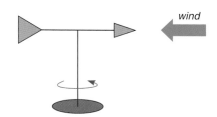

Figure 1.10. A wind vane points toward the direction that the wind is coming from.

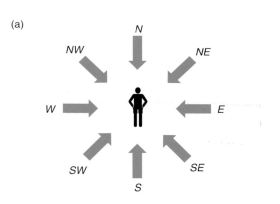

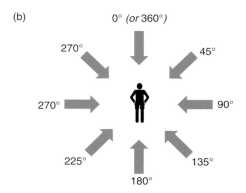

Figure 1.11. Wind direction convention in meteorology. (a) Compass direction. (b) Degrees from true north.

(a)

(b)

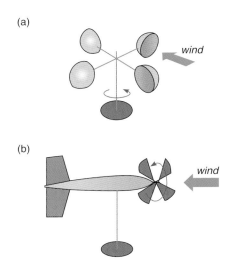

Figure 1.12. Two different instruments to measure wind speed. (a) Cup anemometer. (b) Propeller anemometer.

Table 1.2. *Beaufort wind scale*

Force	Description	Wind speed (knots)	Wind speed (km/h)
0	Calm	0	0
1	Light air	1–3	1–5
2	Light breeze	4–6	6–11
3	Gentle breeze	7–10	12–19
4	Moderate breeze	11–16	20–28
5	Fresh breeze	17–21	29–38
6	Strong breeze	22–27	39–49
7	High wind	28–33	50–61
8	Gale	34–40	62–74
9	Strong gale	41–47	75–88
10	Storm	48–55	89–102
11	Violent storm	56–63	103–117
12	Hurricane force	>63	>117

to the sea state in the absence of anemometers; the rougher the sea surface, the higher the wind speed. It is now also occasionally used on land and describes typical damage observed during strong wind events (Table 1.2).

1.3.2 Reporting Wind

As you know from experience, the wind does not blow constantly and regularly. It stops, resumes, increases, and decreases in strength over short time intervals, producing **wind gusts**. Figure 1.13, for example, shows a lot of "gusty" variability over the course of 2 minutes. Wind gusts are often due to the presence of swirls of air called "turbulent eddies" that mix the air up and down on a variety of scales (i.e., eddies of different sizes). Since wind speed typically increases with altitude and decreases as it approaches the surface, where there is more friction, turbulent eddies tend to bring higher wind speeds down to the surface intermittently (Figure 1.14). The resulting wind gusts can be twice as strong as the average wind and can damage trees, for example, or affect sailing. In the example shown in Figure 1.13, wind gusts reach speeds up to 20 knots.

To reduce the effect of turbulent eddies on wind measurements, we measure the wind away from the surface, at a standard height of 10 m (30 ft). Even at this distance above the surface, the wind speed fluctuates significantly, and an instantaneous wind measurement is not very representative. Therefore, we average the wind speed over a period of time and report the wind as both *sustained wind* and *gusts*, where the gusts denote the higher values of wind speed that only last for short periods of time. The standard time average for sustained surface winds is one minute. In the example shown in Figure 1.13, the sustained wind is about 12 knots, and the wind would be reported as a 12 knot wind with 20 knot gusts.

Note that the effect of friction on the wind varies with the underlying surface. For example, water surfaces are relatively smooth compared to land surfaces, where trees and buildings act as a drag on the air and slow it down. As a result, under similar weather conditions, sustained winds at the surface are typically 30–50% higher over water than over land (Figure 1.15).

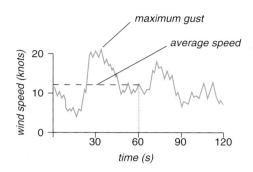

Figure 1.13. Wind variability, wind gusts, and sustained wind (one-minute average wind).

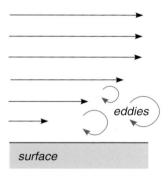

Figure 1.14. Turbulent eddies bring higher wind speeds down to the surface.

1.3.3 Additional Sources of Wind Information

Our understanding of atmospheric motions requires that we measure not only surface winds, but also upper-level winds. Meteorologists use radiosonde measurements to learn about the wind at various altitudes. Wind speed and direction are inferred from the balloon drift, which give us a *vertical wind profile* (see Figure 5.9 for an example). Measurements are also made routinely on **commercial aircraft**, which inform us about the weather along the main carrier routes. Finally, wind measurements can also be obtained from **satellites**, either by tracking cloud movement (speed and direction) or by using radar to measure the small ripples caused by the wind blowing on the surface of the ocean, which is an effect you can also observe on a lake: stronger winds make the surface "rougher."

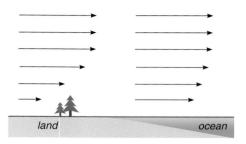

Figure 1.15. The ocean is a smooth surface that offers less resistance to the wind than land surfaces, resulting in higher wind speeds over the ocean.

1.4 Precipitation

The last element of weather that is of particular interest to us is precipitation, a generic term that encompasses rain, snow, hail, sleet, etc. Generally, we can define precipitation as solid or liquid water falling out of the atmosphere. We typically measure precipitation with a **rain gauge**, of which there are two primary types.

In a **tipping bucket**, water is collected by a funnel and falls onto a tipping plate that tips when enough water has accumulated in the bucket (Figure 1.16). One tip usually corresponds to 0.25 mm (0.01 inches) of rain. The number of tips indicates the amount of rain that has fallen over time.

In a simpler type of rain gauge, a **funnel** also collects raindrops and magnifies the rain accumulation into a thinner cylinder so that the total depth may be accurately read on the scale (Figure 1.17). However, only the total depth of accumulated rain is read on the scale. In other words, we have no indication of the changes in precipitation rate between two readings. Such a rain gauge usually measures up to 25 mm, and the overflow is caught in the outside cylinder, for later measurements.

Precipitation of less than 1 mm is reported as a *trace*. Frozen precipitation like snow is melted and its liquid depth is reported. (The frozen depth is also reported.)

On weather maps, precipitation is indicated using specific symbols (see Box 2.1 in Chapter 2). The most useful symbols are those for rain and snow. However, we will discuss very important weather conditions associated with sleet △ (frozen raindrops in the form of ice pellets), freezing rain ⌒⌄ (liquid raindrops that freeze when they hit objects at the ground), and fog ≡ (which is *not* precipitation).

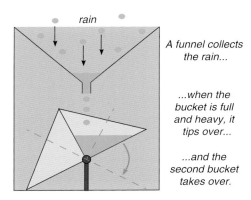

Figure 1.16. Concept of the "tipping bucket" rain gauge.

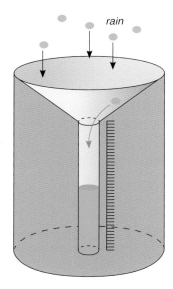

Figure 1.17. Concept of the "funnel" rain gauge.

1.5 Weather Stations

All the instruments described above are set up in a standardized fashion at thousands of weather stations around the world. Most are now automated, as is the case for the Automated Weather Observing Systems (**AWOS**) of the Federal Aviation Administration and the Automated Surface Observing Systems (**ASOS**) of the US National Weather Service (Figure 1.18). At standard sites, thermometers are enclosed in a shelter to prevent direct exposure to the sun, wind, and precipitation. Nevertheless, the shelter needs to be well ventilated so that the measurements are

representative of the surrounding air. That is why weather instrument shelters are painted white (to reflect sunlight) and have open slats to promote air circulation (Figure 1.19). Temperature is measured at a standard height of 2 m.

Figure 1.18. Example of an automated weather station.

Figure 1.19. Standard weather station shelter.

Anemometers are set at the top of "masts" where they can measure wind speed and direction at the standard height of 10 m.

Over the ocean, weather instruments are attached to masts on **buoys**, which are themselves anchored to the sea floor (Figure 1.20). Due to the difficulty and expense of installing and maintaining instruments at sea, there are far fewer buoys than there are land stations, as shown in Figure 1.21. When comparing the number of weather buoys at sea to the number of weather stations on land, it is immediately apparent that weather is much better directly monitored over land, and especially in urban areas, than over the ocean. (Note that some regions have chosen to set up denser networks of weather stations, one of which is apparent in Alberta, Canada.) This disparity has important implications, as weather observations are crucial for constraining forecasting models (see Chapter 13). However, the advent of satellite observations in the last few decades has very much reduced the impact of this disparity, as we will see in Chapter 2.

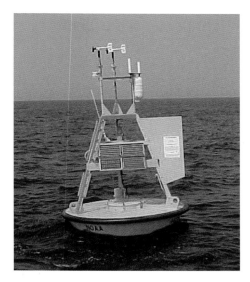

Figure 1.20. Example of a weather buoy.

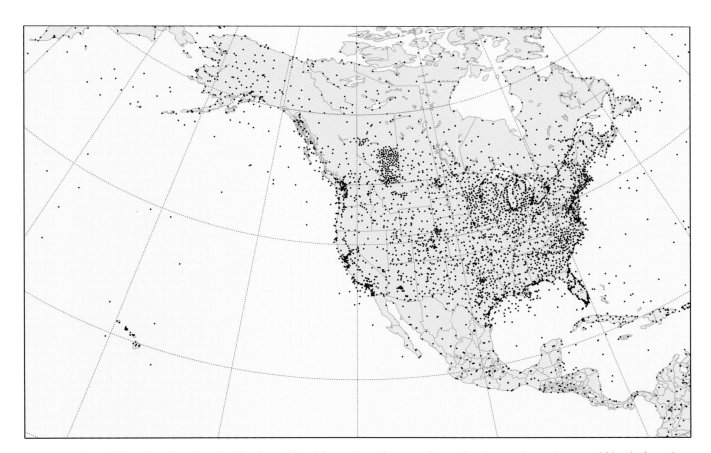

Figure 1.21. Map showing the distribution of land-based stations and weather buoys in and around North America.

Summary

Atmospheric changes obey the laws of physics for the conservation of energy, mass, and momentum. We observe the atmosphere by measuring a number of variables constrained by these laws and relevant to explaining and predicting weather patterns.

▶ We measure **temperature** using thermometers and use three different temperature scales: Celsius, Fahrenheit, and Kelvin.

▶ We measure **atmospheric pressure** with barometers, and express it using hectopascals.

▶ We measure **wind direction** with wind vanes, and **wind speed** with anemometers. We report wind speed as sustained winds and wind gusts, using knots, miles per hour, meters per second, and kilometers per hour.

▶ We measure **precipitation** depth with "funnel" rain gauges with units of millimeters or inches, and precipitation rate with "tipping bucket" rain gauges.

Finally, we assemble weather instruments at dedicated **weather stations** on land using shelters and masts, and buoys at sea. Weather observations are more dense over land than over water.

We can use these measurements to create **time series** of the variables (i.e., variations in time) and **weather maps** (i.e., variations in space at a given time, as we will see in Chapter 2).

Instruments are also attached to a weather balloon, a radiosonde, which records the corresponding variables as the balloon ascends, building a **vertical profile** of the atmosphere above a given location (i.e., variations with height).

Spatial Representations of Weather Data

Local weather is largely the result of large weather systems in motion. Thus, meteorologists gain insight into the manifestations of weather by studying maps and images of the atmosphere on a number of scales, from global, to regional, down to local scales. In this chapter, we will learn how weather information is represented on weather maps and images to reveal the two-dimensional dynamics of weather systems.

In Chapter 1 we introduced some foundational concepts to initiate a study of the atmosphere guided by scientific principles and the laws of physics and, in particular, conservation laws that constrain the behavior of particular variables: temperature, pressure, wind, and water. Understanding weather, and therefore the atmosphere in motion, now requires that we display these variables on maps, to reveal their spatial distribution. Furthermore, because the analysis of raw observations is often quite challenging, we need to design tools that will enable us to analyze these variables in a way that reveals salient features of the weather.

The spatial representation and analysis of data will be the object of this chapter, before we return to each element of weather individually for a more thorough exploration in later chapters.

2.1 The Station Model

The variables described in Chapter 1 and measured at weather stations or buoys are typically displayed on weather maps using the **station model** partly described in Box 2.1. Pressure, temperature, and dew point temperature are displayed as numerical values, while we use symbols for precipitation. The central circle indicates cloud cover. Wind is displayed with wind barbs using the meteorological convention. Wind direction is represented with a line extending from the weather

Box 2.1. A (Simplified) Weather Station Model

Weather observations are reported on surface maps using the weather station model shown in Figure 2.1.1.*

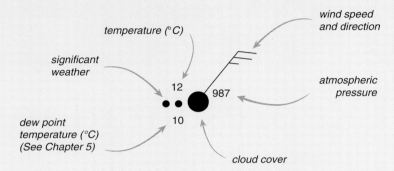

Figure 2.1.1. Simplified weather station model.

Cloud cover is indicated by shading the circle, as shown in Figure 2.1.2.

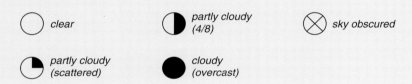

Figure 2.1.2. Some cloud cover symbols.

Significant weather by type and intensity is indicated on weather maps using symbols such as those shown in Figure 2.1.3.

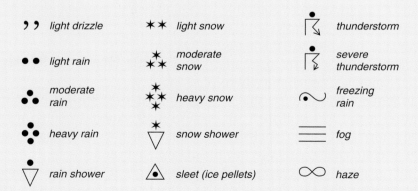

Figure 2.1.3. Some weather symbols.

* See http://www.wpc.ncep.noaa.gov/dailywxmap/wxsymbols.html for a complete description.

Wind direction is represented by a line extending from the weather station toward the direction the wind is *coming from*, while **wind speed** is indicated with barbs and pennants, as shown in Figure 2.1.4.

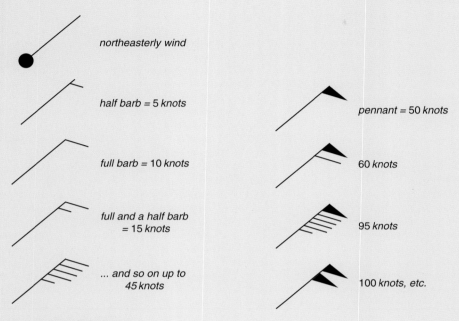

northeasterly wind

half barb = 5 knots

full barb = 10 knots

full and a half barb = 15 knots

... and so on up to 45 knots

pennant = 50 knots

60 knots

95 knots

100 knots, etc.

Figure 2.1.4.   Wind convention.

Strictly speaking, the barbs and pennants indicate a *range* of wind speeds around the central value. For example, a half barb indicates 3 to 7 knots. Hence the values shown in Figure 2.1.5.

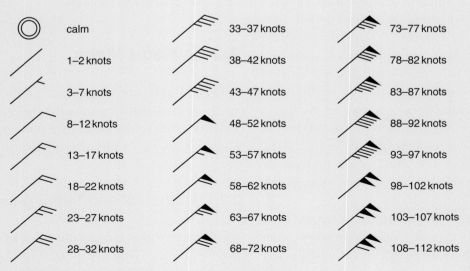

calm

1–2 knots

3–7 knots

8–12 knots

13–17 knots

18–22 knots

23–27 knots

28–32 knots

33–37 knots

38–42 knots

43–47 knots

48–52 knots

53–57 knots

58–62 knots

63–67 knots

68–72 knots

73–77 knots

78–82 knots

83–87 knots

88–92 knots

93–97 knots

98–102 knots

103–107 knots

108–112 knots

Figure 2.1.5.   Reporting wind speed with barbs and pennants.

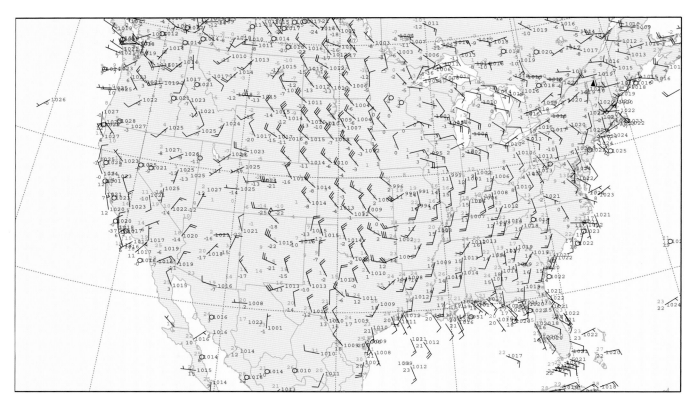

Figure 2.1. Example of surface station map on February 20, 2014, at 18:00 UTC. To facilitate reading, temperature is indicated in red, dew point temperature in blue, and sea-level pressure in black, in addition to following the weather station model layout described in Box 2.1.

station toward the direction the wind is *coming from,* while wind speed is indicated with barbs and pennants.

An example of a surface weather map showing measurements at land and buoy stations on February 20, 2014, is shown in Figure 2.1 (only temperature, dew point temperature, pressure, and wind are shown, for clarity). This is our first look at a weather event that will follow us through the rest of this book: an extratropical cyclone generated east of the Rocky Mountains and sweeping through the eastern half of the United States between February 19 and 22, 2014. Extratropical cyclones are also called midlatitude storms, or frontal cyclones, as cyclones form fronts, but fronts can also form cyclones.

At this stage of its development, the storm (let's call it "our February 2014 cyclone") appears in our weather map as a counterclockwise circulation occurring over the eastern half of the United States. By following the direction of the wind barbs, we can get a sense of its size and locate its center somewhere along the border between Iowa and Missouri (see Appendix 2.2

for geographical reference). By contrast, the western United States is relatively quiet with low wind speeds and no organized circulation.

2.2 Surface Maps

As evident in Figure 2.1, it is quite difficult to appreciate the spatial distribution of the different variables at a glance. That is why we analyze and complement the surface maps with additional elements.

2.2.1 Isotherms and Temperature Maps

We first color the temperature field using an intuitive color scale, typically going from blue (low temperatures) to red (high temperatures), as shown in Figure 2.2. We further contour the temperature observations with contour lines at regular intervals. These contours are called **isotherms**, from "iso," meaning

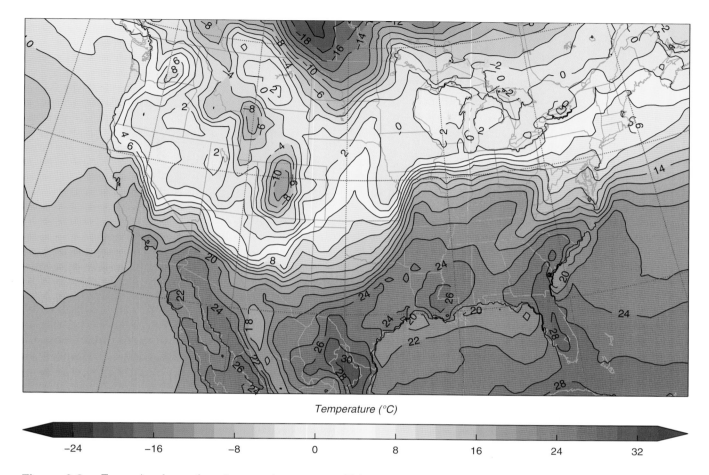

Temperature (°C)

Figure 2.2. Example of a surface temperature map on February 20, 2014 at 18:00 UTC.

"equal," and "therm", meaning "heat." Isotherms are lines of constant temperature. Indeed, if we were to walk along an isotherm with a thermometer, the temperature would remain constant. If we decided to depart from the isotherm, the temperature would either increase or decrease.

Coloring and contouring the temperature map allows us to quickly identify the **air masses**, where the temperature is relatively uniform. In Figure 2.2, for example, a warm air mass occupies the southeastern sector of the United States, while a cold air mass occupies the northwestern half. In between these air masses are areas of strong temperature change, where the temperature increases from cold to warm over relatively short distances. These regions of sharp temperature change are called **temperature fronts** – see Section 2.2.2.

The rate of change of a variable over a certain distance is called a **gradient** – a notion we will use often in the rest of this book. In Figure 2.2, for example,

we can see that the temperature gradient is relatively small in the cold and warm air masses, i.e., if we drove across an air mass with a thermometer, for example from Austin, Texas, to Atlanta, Georgia, a 1500 km drive, the temperature would not vary much. Frontal regions, however, are, by definition, regions of strong temperature gradient, i.e., if we drove across a temperature front, our thermometer would record a strong temperature increase or decrease over a short distance. Going from Kansas to Arkansas, for example, the temperature increases by 30 °F over only 200 miles.

2.2.2 Temperature Fronts

Weather in the midlatitudes is largely the result of air masses in motion. In fact, by matching the air masses visible in Figure 2.2 with the corresponding wind directions in Figure 2.1, you can see that the

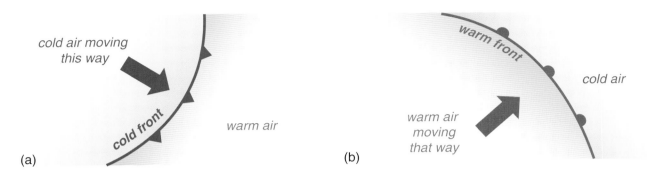

Figure 2.3. Conventional representation of (a) a cold front and (b) a warm front on weather maps.

warm air mass over the southeastern United States is advancing northward, toward colder air, while the cold air mass over the Great Plains is advancing to the southeast, toward the warm air mass. The temperature fronts visible in between are largely the result of these air masses in motion. As we will see in Chapter 10, clouds and precipitation tend to be concentrated along the temperature fronts, i.e., along the boundaries between the air masses. Therefore, we indicate the leading edge of air masses on weather maps, using a blue line and blue triangles for cold fronts, and a red line and red semi-circles for warm fronts, as shown in Figure 2.3. Note that, in each case, the symbols are pointing in the direction of motion. In Figure 2.3(a), the cold air mass is replacing the warm air to the southeast, and therefore the blue triangles point to the southeast. Similarly in Figure 2.3(b), the warm air replaces the cold air to the northeast, and therefore the red semi-circles point to the northeast. [For future reference, know that fronts are drawn on the *warm side* of the largest magnitude in the temperature gradient.]

We will learn more about air masses and fronts in Chapter 10. In the meantime, we will soon add cold and warm fronts to our surface map, but first we need to introduce pressure contours, or isobars, which will help identify the fronts.

2.2.3 Isobars and Pressure Maps

As for temperature, when studying isolated point measurements of pressure on a surface weather map, it is

quite challenging to capture visually at once the entire structure of the pressure field, with its minimum and maximum values, its areas of rapid horizontal change as opposed to slow change, etc. That is why we also contour the pressure observations, to obtain a two-dimensional picture of the pressure field (Figure 2.4). Note that pressure measurements are adjusted to mean sea level before contouring (Box 2.2). Contours of pressure are called **isobars**, and each isobar represents a particular value of pressure – typically at a 4 hPa interval. We can talk, for example, about the 1000 hPa isobar. By definition, the pressure will be 1000 hPa everywhere along that contour.

A good analogy for isobars is the lines of constant height on a topographic map. You know that if you are hiking along such a line, you stay at the same height. If you depart from that contour, however, you either climb up toward higher elevations or down toward lower elevations. Similarly on a pressure map, pressure necessarily increases on one side of the isobar and decreases on the other side. Furthermore, if topographic contours are tightly packed together, you know that you are facing a steep slope. Similarly, tight isobars indicate a steep pressure change, i.e., a sharp pressure *gradient*, and we will learn in Chapter 8 that strong pressure gradients induce strong winds.

2.2.4 Highs, Lows, Ridges, and Troughs

The surface pressure field often contains regions of minimum and maximum pressure. Figure 2.5(a) shows an example of a high pressure region where the

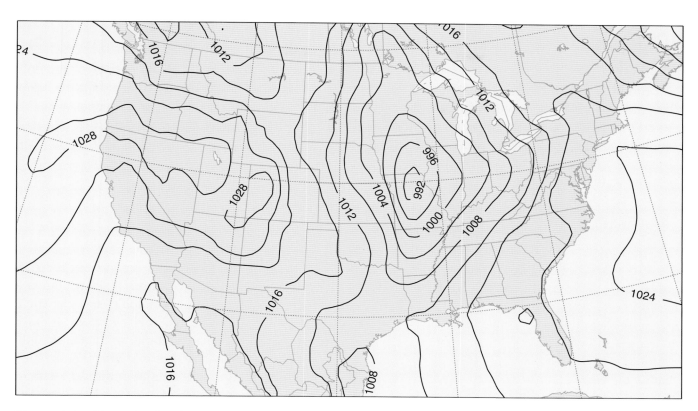

Figure 2.4. Example of surface pressure map on February 20, 2014, at 18:00 UTC.

pressure increases toward a maximum value, which we indicate with a big **H** (for "**high**"). We will see that high pressure regions, or "highs," correspond to anticyclones and are often associated with fair weather (i.e., weak wind and clear skies). In Figure 2.4 the western half of the United States is under a broad area of high pressure.

Figure 2.5(b) shows an example of a low pressure region in which the pressure decreases to a central minimum, indicated with a big **L** (for "**low**"). Note that there is no 988 hPa contour. Therefore, we know that the pressure is above 988 hPa everywhere inside the interior contour. The big **L**, however, tells us that there is a minimum value less than 992 hPa. The exact value of the minimum pressure is often indicated on standard weather maps. We will see that such low pressure centers correspond to cyclonic weather systems, such as midlatitude storms and hurricanes. They are often associated with active weather, such as clouds, precipitation, and strong wind. We will refer to such minima in pressure as "low pressure areas", "low

pressure centers," or simply as "lows." Our February 2014 cyclone, for example, as depicted in Figure 2.4, corresponds to a low pressure center located over northern Missouri.

To continue with the topography analogy, it is sometimes useful to think of a high pressure region as a *mountain* of high pressure, and to think of a low as a *basin*. For practice, try to convince yourself that you could hike up to the top of the mountain shown in Figure 2.5(a) and down into the bowl shown in Figure 2.5(b).

Similarly, if the contours are not closed, we can still find useful features to identify in the pressure field. For example, if higher pressure values extend in one direction, this is like a mountain ridge, as indicated in Figure 2.6(a). Even though there is no isolated high pressure *center*, the crest of the ridge is still a line of relative maximum pressure, compared to the lower values of pressure down the slopes on each side of the crest. We will refer to this feature as a "**ridge** of high pressure," and to the line of relative maximum pressure as the "axis" of the ridge.

Box 2.2.   Adjustment of Pressure to Mean Sea Level

Pressure maps are an important tool of the weather analyst, as pressure is intimately related to wind and weather systems. We will learn in Chapter 8 that wind is caused by horizontal pressure variations. Pressure, however, decreases far more rapidly upward than it changes horizontally. A good rule of thumb is that pressure decreases by about 8 hPa with each 60 m gain in altitude (at low elevation). Thus, the pressure at the top of the Empire State Building (381 m) is about 50 hPa lower than at street level. The pressure at the top of the Rocky Mountains could be as low as 600 hPa. However, the horizontal pressure variations causing wind are much more subtle than this. Recall that the typical range of pressure at sea level is about 980 to 1030 hPa. Therefore, if we reported actual pressure measurements on a surface weather map, as they are measured at weather stations on the ground, we would observe very low pressure values wherever there are mountains. We would suggest the presence of artificially low pressure areas where the mountains stand, with a drop in pressure of 300 or 400 hPa in places.

To avoid such artefacts, we calculate what the pressure *would be* if there were no mountains and we were at sea level. We *replace* the mountains with air, so to speak, and add the corresponding pressure to the value measured at the weather station. In other words, we *adjust* the values of pressure to mean sea level. Figure 2.2.1(b) shows the *un*adjusted pressure field, and a comparison with the topography (Figure 2.2.1(c)) confirms that the low pressure areas correspond to the Rocky Mountains and the Appalachians, at the expense of our February 2014 extratropical cyclone over the Midwest. After adjusting the pressure measurements to mean sea level (Figure 2.2.1(a)), the low pressure center and the troughs corresponding to the cold and warm fronts appear clearly. [Note that we have used a green color scale for the purpose of this exercise, although pressure fields are usually displayed with isobars alone.]

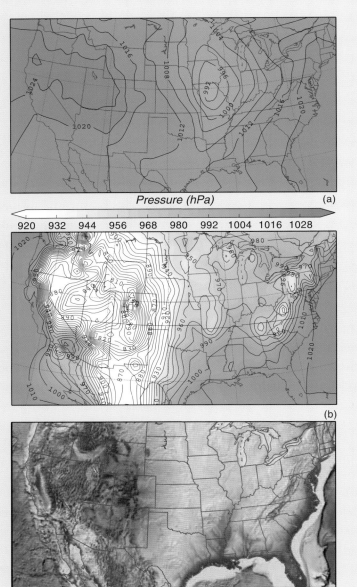

Pressure (hPa)   (a)

920   932   944   956   968   980   992   1004   1016   1028

(b)

(c)

Figure 2.2.1.   Pressure measurements adjusted to mean sea level (a), with the unadjusted pressure field (b) and topography (c) for comparison.

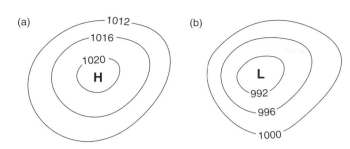

Figure 2.5. Example of (a) high pressure region, or "high," and (b) low pressure center, or "low," as would appear on a surface pressure map.

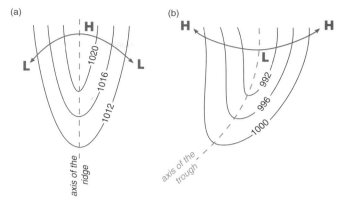

Figure 2.6. Example of (a) high pressure ridge, or "ridge," and (b) low pressure trough, or "trough," as would appear on a surface pressure map.

We will see that ridges are especially apparent on upper-level maps, and are associated with fewer clouds and lighter winds. Conversely, when the pressure field is relatively lower along a line, we have the equivalent of a *valley*. Although we do not call it a "valley of pressure," we use a similar analogy and call it a "**trough of low pressure**." Even though there is not a closed low pressure center, the axis of the trough is still a line of relative minimum pressure (Figure 2.6(b)). We will see that troughs are also frequently observed on upper-level maps and are often associated with regions of clouds and precipitation.

Note that we rarely *color* the pressure field as we did for temperature, because it is more informative to overlay the isobars on top of the (colored) temperature field, as shown in Figure 2.7. We will see that this is a powerful tool in meteorology, as it allows us to analyze temperature and pressure jointly. After adding the highs (**H**), the lows (**L**), and the temperature fronts, we obtain a compelling picture of the state of the atmosphere at any given time. To convince yourself fully, return to Figure 2.1 and compare it to Figure 2.7 to appreciate how much more readable our surface map has become. At a glance, we have what we call a **synoptic** view of the atmosphere, or a view of the "synoptic situation" over the United States (from the Greek "syn" meaning "together" and "opsis" meaning "view").

Here we can see that the cold and warm air masses, the cold and warm fronts, and the low pressure center all coincide to form our February 2014 extratropical cyclone. We will see in Chapter 10 that the air masses and fronts circle around the low, and the **warm sector**,

i.e., the sector of warm air delimited by the two fronts, decreases in size as the cyclone evolves, while clouds and rain form along the fronts and at the center of the cyclone. By looking carefully at the pressure field, you will also notice that the fronts happen to be located where the isobars "kink" and change directions. If you think of these two depressed areas as pressure *troughs*, then the fronts follow the axis of the troughs. This is indeed a useful tip for placing temperature fronts on a surface map, the reason for which we will discover when we know a bit more about wind.

2.3 Upper-Level Maps

We tend to think of weather as happening close to Earth's surface, where we live and where we can observe it. But most of the action actually takes place aloft. Surface weather in the midlatitudes is largely determined by the position and structure of the jet stream, a fast-flowing air current blowing at the top of the troposphere. Disturbances in the jet stream create certain conditions that are more or less conducive to the formation, or demise, of cyclones. That is why meteorologists spend a lot of time studying upper-level maps.

We could create such maps by contouring pressure observations at specified heights, as we did for our surface weather map. For example, we could collect pressure measurements at an altitude of 5000 m and

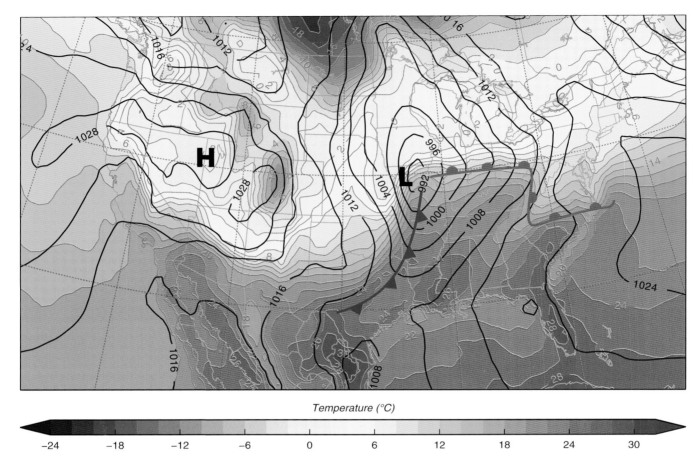

Temperature (°C)

−24 −18 −12 −6 0 6 12 18 24 30

Figure 2.7. Example of surface weather map showing temperature (colors and isotherms), pressure (isobars), highs (**H**) and lows (**L**), as well as temperature fronts on February 20, 2014, at 18:00 UTC.

build a 5000 meter pressure map. We would then find highs and lows, troughs and ridges, as we did at the surface. In practice, however, it is more convenient to do the opposite: we choose a particular pressure, for example, 500 hPa, and we indicate how high in the atmosphere this 500 hPa value is located. If we now contour these height measurements, we have a 500 hPa "height" map. We can think of it as a two-dimensional surface where the pressure is everywhere 500 hPa, roughly at 5.5 km altitude, but with dips and bumps that make it sag lower or rise higher in places.

This seems complicated, but in practice it yields about the same result, and it is more convenient from a mathematical point of view, because working at constant pressure allows us to simplify our equations.

[Specifically, the air density disappears from the equations. Recall that density is not measured, but estimated indirectly from other measurements. Therefore, disposing of density is very convenient.]

In the end, we can treat this "height" map exactly as we would a pressure map, as illustrated in Figure 2.8. If you picture a series of pressure surfaces at different altitudes, they look like the blue lines shown in Figure 2.8(a) on a cross-section. Recall that pressure decreases with height, and therefore the 850 hPa surface is necessarily *above* the 900 hPa surface and *below* the 800 hPa surface. If you follow the 850 hPa surface from west to east, you will see that it is higher to the west (blue **H**) and lower to the east (blue **L**). If you now follow the 1.5 km height line (dashed) from west to east and read the corresponding pressure (on the

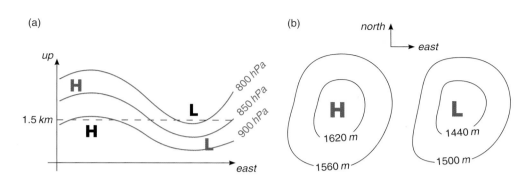

Figure 2.8. Example of a high and a low as would appear on (a) a cross-section, and (b) a 850 hPa map. The blue lines in panel (a) represent pressure surfaces, while the black lines in panel (b) represent isobars. The blue **H** represents a region where the 850 hPa surface is high, while the black **H** represents the corresponding region where the pressure is high at 1.5 km altitude (and similarly for low pressures).

blue lines), you will see that the pressure is higher to the west (about 900 hPa, indicated by a black **H**) and lower to the east (about 800 hPa, indicated by a black **L**). In other words, the height surface is high (**H**) when the pressure is high (**H**), and it is low (**L**) when the pressure is low (**L**). On the actual 850 hPa map, the high and the low might look like Figure 2.8(b), which tells us that the 850 hPa surface is sloping from higher to lower altitudes from **H** to **L**. The high (at the constant pressure of 850 hPa) also corresponds to higher pressures (at the constant height of 1.5 km), while the low corresponds to lower pressures. Therefore, for our purposes, we can think about the features on the upper-level height maps as if they were on a pressure map.

Meteorologists frequently use the 850, 500, and 300 hPa pressure surfaces for upper-level analyses. In addition to height contours, we usually plot station data on these maps, as shown in Figure 2.9 for a 500 hPa map, which helps us to visualize the speed of air currents flowing along the contours (see Chapter 8). Finally, recall from the ideal gas law that, for a fixed mass of air, warmer air has a greater volume than colder air. Therefore, for a fixed area, a warmer column of air is "thicker" and associated with higher upper-level heights on pressure surfaces. Thus, meteorologists can also analyze height fields in terms of thickness and heat content of the underlying air columns.

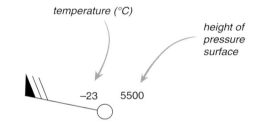

Figure 2.9. Example of station data on a 500 hPa upper-level map. Height given in meters.

An example of a simplified upper-level map, corresponding to our February 2014 cyclone, is shown in Figure 2.10, where wind speed is also indicated with colors in the background. Closed highs and lows are not so common at 500 hPa, but troughs and ridges are, and often correspond to specific weather features at the surface. Upper-level ridges, for example, often indicate fewer clouds, while upper-level troughs create surface conditions that are conducive to the formation of cyclones (see Chapter 10). Here, the Great Lakes are under a ridge of high pressure, but the Midwest is under a trough of low pressure, with strong winds blowing around the trough, and particularly high winds on the southeastern flank of the trough, over Missouri. The alert reader will notice that this feature of the upper-level jet coincides with the location of the surface low in Figure 2.7. Indeed, we will

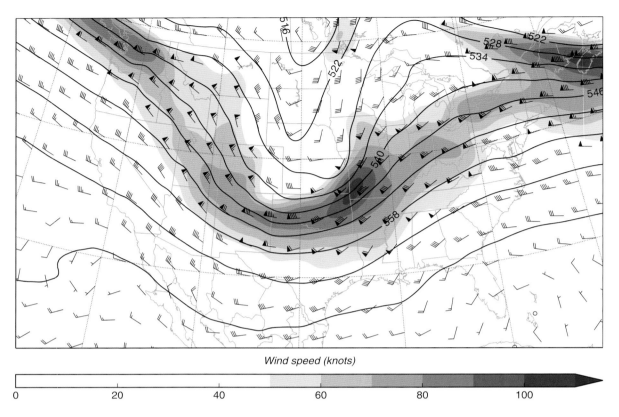

Wind speed (knots)

| 0 | 20 | 40 | 60 | 80 | 100 |

Figure 2.10. Example of a 500 hPa map on February 20, 2014, at 18:00 UTC.

learn in Chapter 10 that this configuration is conducive to the deepening of extratropical cyclones.

2.4 Radar

Radars (originally RAdio Detection And Ranging when developed for military purposes) are very useful for measuring and tracking precipitation. They emit radio waves nearly horizontally, since the troposphere is so thin. Some of these waves are transmitted through the atmosphere and are lost if they do not encounter any obstacle. If precipitation is present on their path, however, the raindrops will reflect some of the waves back to the radar, where the returned energy can be measured and analyzed (Figure 2.11).

The return signal is an indirect measurement of precipitation. It contains two pieces of information. First, more precipitation causes more reflected signal. Therefore, the radar can be used to estimate the *rate* of precipitation. Second, by calculating the time it took for the signal to reach the raindrops and

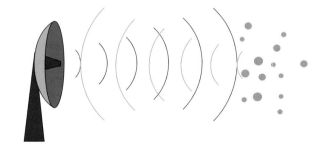

Figure 2.11. Concept of radar as used to measure precipitation. The black lines show the signal emitted by the radar. The red lines show the part of the signal that returns to the radar after being reflected by raindrops.

come back, we can estimate the *distance* between the radar and the precipitation, and we can indeed *map* the structure of the precipitation around the radar. Therefore, a radar provides information about both the location and the intensity of precipitation.

By tilting the radar up and down, we can also learn about the vertical structure of the precipitation. By rotating the radar horizontally, we can build a 360° picture

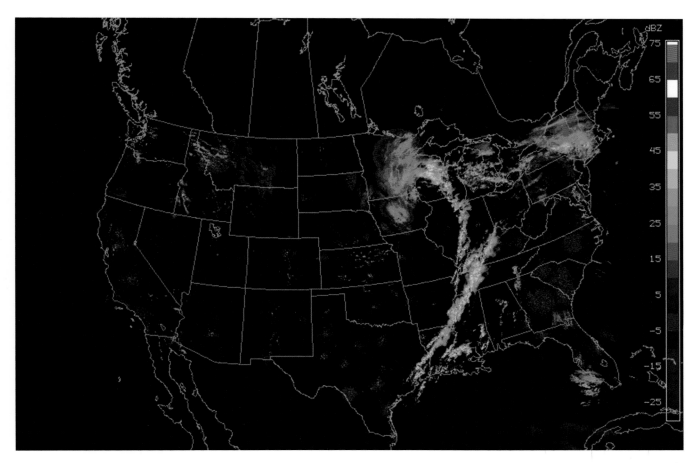

Figure 2.12.   Example of a radar image of precipitation over the United States on February 21, 2014, at 00:00 UTC.

of precipitation all around the radar, out to a distance of about 250 km. By aggregating measurements from a network of radars, we can build a composite picture for large areas, as shown in Figure 2.12 for our February 2014 cyclone at a later stage of its development. Recall that precipitation tends to concentrate along the temperature fronts and in the low pressure center. On February 21, the cyclone center had moved over Wisconsin, and the bands of precipitation can be seen to delineate the cold front (from East Texas to Illinois), the warm front (from Michigan to New York), and what we will call the occluded front (arching back over Wisconsin).

A more advanced type of radar, called **Doppler radar**, can also detect the shift in frequency of the return signal, as compared to the emitted signal, caused by the horizontal motion of the raindrops. This shift in frequency can be translated into a speed, which informs us about the wind – specifically, whether the raindrops are moving toward or away from the radar.

Finally, the latest radars have the ability to utilize polarized radar signals, i.e., radio waves that are oriented in the vertical and horizontal plane, to determine the shape of precipitation particles.

2.5 Satellites

Since the first weather satellite was placed in space in 1960 – TIROS (Television Infrared Observation Satellite) – satellites have revolutionized our view and understanding of the atmosphere and weather. They provide information about cloud coverage (cloud location and height), as well as information about the structure of weather systems. Importantly, they provide information both over land and over the ocean, where very few direct measurements are available. (Recall that the ocean covers two thirds of the planet.)

Figure 2.13. Example of a visible satellite image on February 21, 2014, at 17:45 UTC. For orientation, the United States and Mexico appear in light gray in the western half of the image.

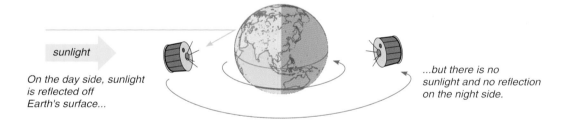

sunlight

On the day side, sunlight is reflected off Earth's surface...

...but there is no sunlight and no reflection on the night side.

Figure 2.14. Geostationary satellites can only take visible images during the day. A few hours later, both Earth and the satellite have spun away to the night side.

Weather satellites carry sensors that measure different kinds of radiation. These measurements can then be translated into variables of interest for weather and climate (temperature, cloud coverage, cloud top height, cloud motion, wind, etc.). For our purposes, we will be interested in three important types of satellite sensor: visible, infrared, and water vapor.

2.5.1 Visible Satellite Images

A satellite can be equipped with a visible light sensor. Since Earth does not emit visible radiation, the only light that will be measured by such a sensor is visible light from the sun reflected by bright features on Earth, such as clouds, ice, snow, and to some extent bright land areas. Figure 2.13 shows a visible image of our February 2014 cyclone captured by GOES-East, an American geostationary satellite (see Section 2.5.4). Later on February 21, the cyclone was starting to decay, as we will explain in Chapter 10. At this fully developed stage, however, the visible image reveals a beautiful display of clouds spanning thousands of miles. The cold front has now moved further east and extends from the Gulf of Mexico, along the East Coast, to New England. The extensive cloud cover over the low pressure area suggests precipitation over Canada, and it was

Figure 2.15.   Example of an infrared satellite image on February 21, 2014, at 17:45 UTC, corresponding to Figure 2.13.

indeed snowing heavily over the Great Lakes. As for the warm front, it extends over the Atlantic Ocean and connects with another cyclone in its decaying phase over Europe.

Visible images are useful for identifying and tracking cloud features when they are illuminated by the sun. As Earth and the satellite rotate away from the sun to the night side, however, no more sunlight is reflected and no observation is possible (Figure 2.14). Therefore, we resort to a second type of sensor to obtain information about Earth independently of the position of the sun.

2.5.2 Infrared Satellite Images

As we will see in Chapter 4, all objects emit some level of radiation, and the warmer the object, the more radiation it emits and at a higher frequency. The sun, for example, emits a *large* amount of radiation, in the **visible** part of the spectrum. Most objects on Earth are not as hot, however, and therefore emit a *lesser* amount of radiation, in the **infrared** (IR) part of the spectrum – radiation of lower frequency than visible light. Now,

the infrared spectrum encompasses a broad range of frequencies and slight variations in the temperature of Earth features will lead to slight variations in the amount and frequency of the infrared radiation they emit. An infrared sensor on a satellite can be tuned to detect such variations in radiation, and therefore variations in temperature. Thus, the full picture taken by an infrared sensor, i.e., an infrared satellite image, is really a "temperature image" of Earth.

It is relatively straightforward to display a *visible* satellite image, since it shows in shades of gray what our eyes would see if we were sitting on that satellite looking down at Earth. A visible satellite image is really a black and white photograph, as can be seen in Figure 2.14 – or, lately, a color photograph, with the latest full-color satellite imagery. But how do we display an *infrared* satellite image? Our eyes have evolved to see visible light, but not infrared radiation. Therefore, there is no such thing as "infrared colors." So we need to create a "fake" color scale to display the temperature information contained in the image (Figure 2.15). We display warm features with darker shades of gray and colder features with lighter shades of gray (Figure 2.16). Of course, the scale was

cold (white)

cool (light gray)

warm (dark gray)

hot (black)

Figure 2.16. Color scale for displaying satellite infrared images.

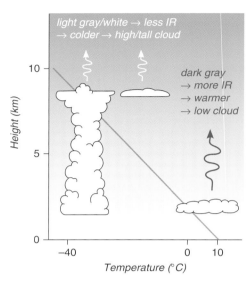

Figure 2.17. The amount of infrared radiation emitted by cloud tops tells us their temperature, which in turn tells us their altitude. (Note that "light gray" and "dark gray" here refer to the appearance of cloud tops on an infrared image, *not* to the color of the clouds themselves.)

not chosen randomly: it reflects the fact that cloud tops, being higher in the atmosphere, are colder than the surface. Recall that temperature decreases with height. Since it is more intuitive to represent cloud tops in white, we chose our scale to show cold as white, and warm as dark gray.

At first glance, the white features in Figure 2.15 indeed seem to match the clouds of Figure 2.13. After closer inspection, however, we notice that some of the clouds present in the visible image are absent from the infrared image. Indeed, clouds are not all found at the same altitude, and therefore do not all have the same temperature. Clouds with a light gray shading in an infrared satellite image are colder and therefore higher in the atmosphere, while clouds with a dark gray shading are warmer and therefore lower in the atmosphere (Figure 2.17). [Note that, strictly speaking, what we are seeing are the *cloud tops*. Therefore, a cloud with a light gray shading can be either a *high* cloud or a *tall* cloud.]

Figure 2.15 tells us that, at this later stage of its development, our February 2014 extratropical cyclone contains different types of clouds, in contrast to what seems to be fairly homogeneous cloud cover in the visible image. This is very instructive to the weather analyst, as it reveals different aspects of the circulation pattern. The cloud band along the cold front, for example, is bright white and reaches high in the atmosphere. In contrast, the cloud spiral that extends over Canada and arches back toward the Great Lakes, producing heavy snow, is darker in color, and therefore lower in the atmosphere.

In the end, we should never forget that an infrared image shows us temperature, rather than the brightness of the clouds. This explains, for example, why some clouds over the Atlantic Ocean appear in the visible image shown in Figure 2.13, but not in the infrared image shown in Figure 2.15. They are clouds at very low altitude. They are made of liquid cloud droplets and highly reflective; therefore, they clearly appear on a visible image. However, being low in the atmosphere, and therefore at temperatures close to the temperature of the ocean underneath, they are hardly distinguishable from the ocean itself. If we were not aware of the true nature of an infrared image, we could simply overlook the existence of these clouds.

2.5.3 Water Vapor Images

Our satellite sensor can also be tuned to detect radiation emitted by specific molecules in the atmosphere. As we will see in Chapter 4, gas molecules absorb and emit radiation at specific wavelengths due to their molecular structure. A gas of particular

Figure 2.18. Example of a water vapor satellite image on February 21, 2014, at 17:45 UTC, corresponding to Figure 2.13.

interest to us is water vapor, as it transports energy through phase changes and plays a major role in the development of weather systems (see Chapter 5). Therefore, we also equip our satellite with a water vapor sensor to produce images such as shown in Figure 2.18.

In this image, darker colors indicate a drier troposphere, whereas lighter colors indicate a higher water vapor content. As for infrared imagery, the altitude of the atmospheric water vapor is a factor in the amount of radiation recorded by the satellite. In particular, dark and light colors are more indicative of the presence or absence of water vapor in the *mid to upper* troposphere. In other words, a dark region on a water vapor satellite image might still contain a fair amount of water vapor in the *lower* troposphere, and even low clouds. Note, for example, the presence of numerous low cumulus clouds over the tropical Atlantic in Figure 2.13, even though it is depicted as "dry" in Figure 2.18.

Water vapor satellite imagery is particularly useful to weather analysts when it is animated, as it shows the fluxes of water vapor in and out of weather systems, as well as the intrusions of dry air. Note, for example, the tongue of dry upper-tropospheric air wrapping around and into the cyclone in Figure 2.18 (west of the cold front). This highlights another important attribute of the water vapor image: it shows us structure in the atmosphere even where there are no clouds.

2.5.4 Geostationary Satellites

Meteorologists rely primarily on two types of satellite: geostationary and polar-orbiting. As their name indicates, geostationary satellites are stationary with respect to Earth ("geo-"). More precisely, they are stationary with respect to a specific location on Earth, so that they constantly observe that same location over time. Since Earth rotates, the satellite needs to move at the exact same speed, otherwise it will start leading or lagging behind the location it is supposed to observe. More exactly, it needs to orbit Earth at the same *angular speed*, i.e., it needs to describe the same *angle* over time. At that speed, the satellite will experience a centrifugal force that will tend to pull

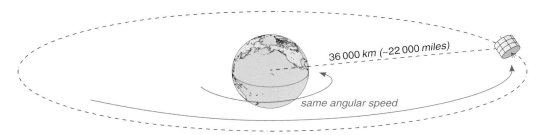

Figure 2.19. Geostationary satellites orbit Earth at the same angular speed as Earth.

it out of its orbit. This is similar to the outward pull you experience when you turn at high speed with your car. For the satellite to stay in orbit, the gravitational pull of Earth needs to balance the centrifugal force exactly. This perfect combination happens in the equatorial plane at about 36 000 km from Earth's center (about 22 000 miles), which defines the **geostationary** orbit (see Figure 2.19). For reference, Earth's radius is about 6400 km. So, a geostationary satellite is at about five to six times that distance from Earth, above the equator. At this location, it can hover above a point on Earth, without falling toward Earth.

Since a geostationary satellite always remains above the same location, it is useful for tracking weather changes over a specific country or region. In the United States, for example, one satellite called GOES-West (Geostationary Operational Environmental Satellite-West, currently GOES-15) covers the western USA and part of the Pacific Ocean, while a second satellite called GOES-East (currently GOES-16) covers the eastern USA and part of the Atlantic Ocean. The full-disk GOES-East visible, infrared, and water vapor images corresponding to Figures 2.13, 2.15, and 2.18 are shown in Figure 2.20.

To obtain a full view of Earth, we need several satellites orbiting Earth in concert, each taking snapshots from a different vantage point. We can do so with a constellation of satellites, as shown in Figure 2.21 with a view of Earth above the North Pole. Europe and Africa are currently covered by two Meteosat satellites, India and Central Asia by INSAT (Indian National Satellite System) and a number of Russian, European, and Chinese satellites, while Japan, Eastern Asia, and Australia are covered by MTSAT (Multifunctional Transport Satellite). Since Earth and the satellites rotate at the same speed, you can turn this book to

model their motion over time. If you turn it counter-clockwise by 90°, you will obtain their new position, as well as that of Earth, in six hours.

The images obtained by the different satellites can be composited to produce a **global** image such as the one shown in Figure 2.22. These global images are very instructive, as they reveal many aspects of the general circulation of the atmosphere, to which we will return in Chapter 9. In particular, it appears that some weather patterns are characteristic of specific latitudinal bands. At the **equator**, for example, the patchy clouds correspond to convective systems due to the convergence of surface winds (see Chapter 6). North and south of the equator, we can see two bands that are relatively free of clouds, at the latitude of the Sahara desert in the northern hemisphere and at the latitude of Australia in the southern hemisphere. These regions are called the **subtropics** and are characterized by subsidence (i.e., downward motion), which prevents the formation of clouds (see Chapter 9). Farther north and south, at the latitude of the United States and the Southern Ocean, we find the **midlatitudes**, i.e., the "middle" latitudes, halfway between the equator and the poles. That is where we find our February 2014 midlatitude cyclone, along with four other cyclones in the northern hemisphere, while eight to ten midlatitude cyclones can be observed over the Southern Ocean at different stages of development (see Chapter 10). They can be recognized by their *comma* cloud signature. Northern and southern midlatitude cyclones are mirror images of each other, for reasons we will explain in Chapter 8. Because the midlatitudes are "outside" of the tropics, i.e., north and south of the tropics, we sometimes refer to them as the **extratropics**, and to midlatitude cyclones as **extratropical cyclones**.

(a)

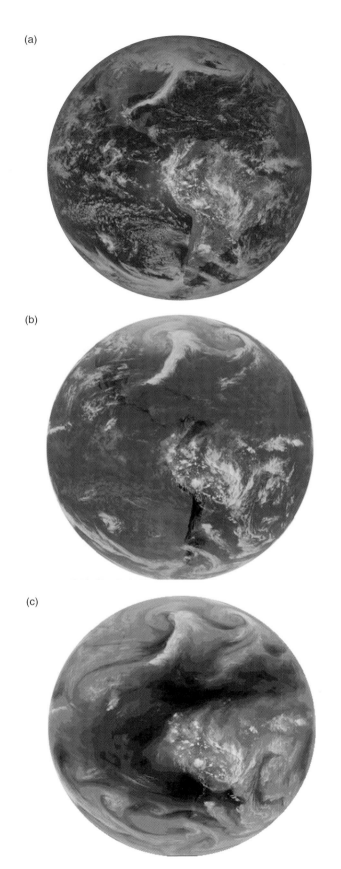

(b)

(c)

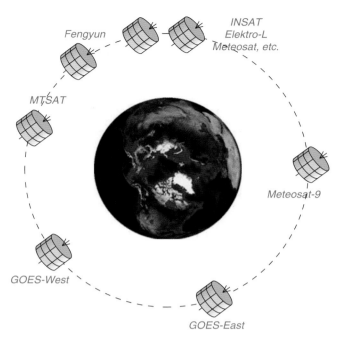

Figure 2.21. Polar view of Earth (looking down at the North Pole) showing the constellation of geostationary satellites observing Earth from all directions.

2.5.5 Polar-Orbiting Satellites

Geostationary satellites are convenient to monitor the weather at a specific location. However, they need to do so from quite far away (36 000 km) and their sensors cannot detect small clouds or small features of Earth's surface. Much as a digital camera can only zoom up to a point, and produces pixelated images beyond that point, satellite sensors can only detect features down to a certain size. Their current resolution is 0.5 km for visible images and 2 km for infrared images, which means that any detail inside a 0.5 km box will be averaged into a single pixel in a visible image. The small cumulus clouds shown schematically in Figure 2.23, for example, are not detectable and will blend into a white or gray pixel in the visible image.

To gain in resolution, we need to bring our satellite closer to Earth. But this has a cost: the gravitational

Figure 2.20. Example of full-disk visible (a), infrared (b), and water vapor (c) images taken by GOES-East on February 21, 2014, at 17:45 UTC.

Figure 2.22. Global satellite image obtained by compositing several images from different satellites on February 20, 2014, at 18:00 UTC.

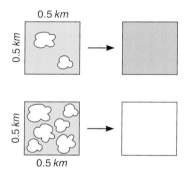

Figure 2.23. Features smaller than 0.5 km blend into a unique pixel of homogeneous color in a visible satellite image.

pull will be stronger, and the satellite will need to go faster to generate enough centrifugal force to balance gravity and remain in orbit. That is the choice we make with **polar-orbiting satellites**. They orbit Earth at about 800 km (one eighth of Earth's radius), which allows them to resolve details of Earth's surface down to about 250 m, but they complete an orbit in about 100 minutes, which prevents them from staying above the same location. They orbit Earth in a pole-to-pole direction, however, and provide valuable information about the polar regions, which are not captured by geostationary satellites. The main disadvantage of the polar-orbiting satellite is that, being so close to Earth, it cannot get a full view of the planet, but only a narrow swath along its path. While it completes one orbit, however, Earth rotates eastward, which means that successive swaths sample different regions of Earth (Figure 2.24). By compositing successive swaths, we can obtain a full picture of Earth in about 24 hours – with some gaps (Figure 2.25). When combining both ascending and descending swaths, we can obtain a picture of Earth in 12 to 14 hours.

Figure 2.25 shows a number of comma-shaped cloud structures typical of extratropical cyclones, spiraling in opposite directions, such as we saw in Figure 2.22. It also features a number of isolated clouds at the equator – the signature of convective systems. In the northern hemisphere, the cloud-free

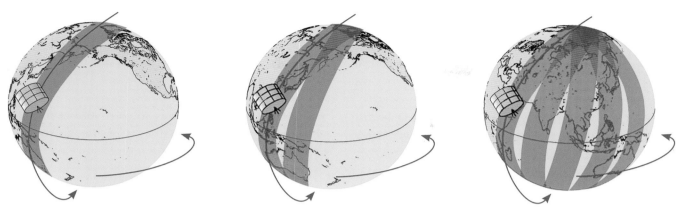

Figure 2.24. Polar-orbiting satellites orbit Earth in about 100 min at an elevation of about 800 km. As Earth rotates underneath, the satellite samples a new region of Earth in each orbit.

Figure 2.25. A polar-orbiting satellite covers Earth in about 24 hours, with minor gaps between the successive swaths, as illustrated with this image from the Moderate-resolution Imaging Spectroradiometer (MODIS) aboard NASA's Terra satellite on April 19, 2000.

subtropical band extending from Mexico to North Africa, the Arab Peninsula, and India, is particularly well delineated. We will explain the existence of all these features in time, as we learn more about the general circulation of the atmosphere and individual weather systems. In the meantime, Appendix 2.1 shows some examples of useful cloud shapes to memorize.

Summary

We depict weather patterns by creating a variety of maps and images.

- **Surface station maps** contain measurements taken at individual weather stations and buoys using a conventional station model.

- **Temperature maps** show the structure of the temperature field using colors and contours called **isotherms**. They are useful for identifying air masses and fronts. **Temperature fronts** appear as regions of larger temperature gradients, characterized by a concentration of isotherms over short distances.

- **Pressure maps** reveal the structure of the atmospheric pressure field using contours called **isobars**. They are useful for identifying regions of high and low pressure, either closed (i.e., **highs** and **lows**), or open (**ridges** and **troughs**). High pressure often corresponds to fair weather, while lower pressures often correspond to clouds and precipitation.

- Troughs and ridges also appear on **upper-level maps**, where we display the height of a given pressure surface, rather than the pressure at a given height (for the same overall interpretation).

- **Radar images** show us the position and intensity of precipitating particles. Doppler radar images show us the speed of the precipitation particles relative to the radar.

- **Visible satellite images** show us the visible light from the sun that is reflected, more or less efficiently, by the various elements of the Earth system, most importantly by clouds.

- **Infrared satellite images** show us the temperature of the various elements of the Earth system, and in particular the temperature of cloud tops. Since temperature typically decreases with height over the depth of the troposphere, this information can be used to infer the altitude of cloud tops, and therefore the type of clouds and the structure of weather systems.

- **Water vapor satellite images** show us the water vapor content of the middle and upper troposphere, and are useful for identifying the fluxes of water vapor and dry air.

- **Geostationary satellites** orbit Earth at about 36 000 km above the equator, where their angular velocity is the same as that of Earth. They can observe the same location over time, but with less resolution than **polar-orbiting satellites**, which orbit Earth at about 800 km. These satellites, however, observe the planet along successive swaths that cover different areas of Earth at different times.

Let's get started

Even though our meteorologist toolkit is not complete, it is sufficient to start investigating the atmosphere and begin learning about the mechanisms of weather. With our basic tools in hand, let's get started on our journey, and lay down the first building blocks of our understanding.

Appendix 2.1　Important Satellite Cloud Signatures

Satellite images allow us to track the cloud signature of various weather systems, such as fronts, cyclones, hurricanes, and thunderstorms. A developing wave, for example, in the first stage of cyclone development, will appear with two levels of clouds (Figure 2.26).

As the cyclone matures, the fronts become more substantial. Cold fronts are usually associated with long, narrow bands of clouds, whereas warm fronts are associated with more extensive cloud decks (Figure 2.27).

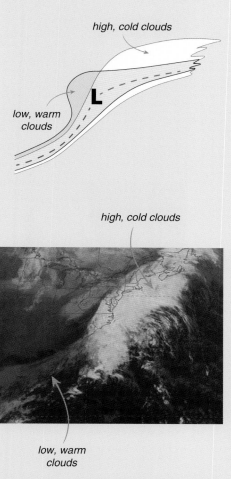

Figure 2.26.　Developing wave cyclone, with two levels of cloud.

Figure 2.27.　Cloud signature of cold and warm fronts.

Low pressure systems are often associated with a cloud swirl spinning counterclockwise (in the northern hemisphere) around the low (Figure 2.28).

Cold fronts can spawn strong convection in the form of individual, growing convective cells (Figure 2.29).

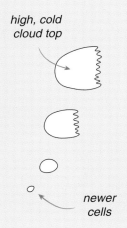

high, cold cloud top

newer cells

Figure 2.28. Cloud structure around a midlatitude low pressure system.

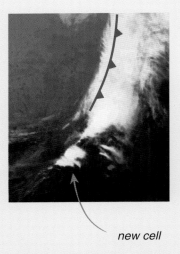

new cell

Figure 2.29. Convective cells in satellite imagery.

Appendix 2.2   Contiguous USA Reference Map

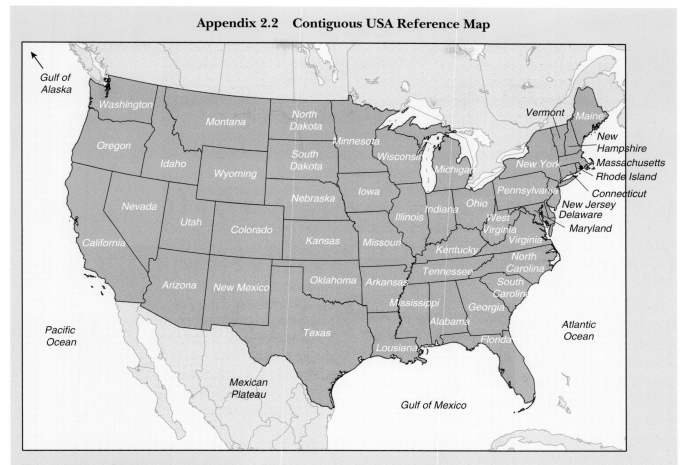

Figure 2.30.   Contiguous states of the USA and surrounding geography.

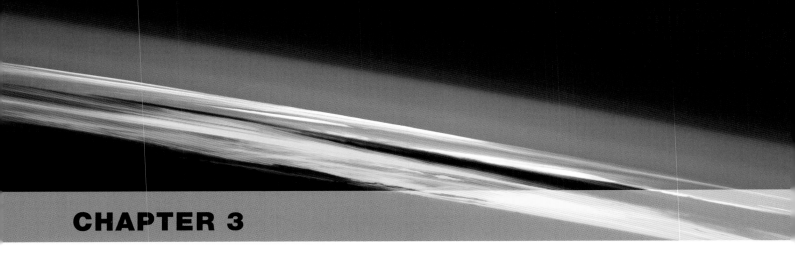

CHAPTER 3

Our Atmosphere: Origin, Composition, and Structure

The atmosphere is a thin layer of gas and particles with a well-defined vertical structure. Weather systems continually stir the lowest layer of the atmosphere, and we experience the clouds, precipitation, and wind that they produce. Before we investigate the formation and evolution of these weather systems, let us look deeper into the nature and composition of the atmosphere, as well as into its origin and future evolution. We will also examine the vertical structure and properties of different layers of the atmosphere.

3.1 Aspect

We can think of the atmosphere as a thin layer of gas surrounding Earth. It is attracted to Earth by gravity and is extremely thin: 99% of the atmosphere is found below 30 km while, by comparison, Earth's radius is about 6380 km. A common analogy compares the thickness of the atmosphere around Earth to the thickness of the skin around an apple. This very thin aspect of the atmosphere has important implications: since large weather systems such as midlatitude cyclones and storms can be several thousand miles wide but less than 10 miles high, we need to think of them as essentially "flat" with a large horizontal extent. This is not to say that vertical air motion is unimportant in

storms and hurricanes – in fact, it is precisely vertical air motion that determines the formation of clouds and precipitation. But it is useful to remember that air masses are mostly flat and sliding around near the surface of Earth.

3.2 Composition

Besides some ions and larger solid and liquid particles, the atmosphere is made primarily of gas molecules. The atmosphere mostly contains molecular nitrogen, N_2 (78%) and molecular oxygen, O_2 (21%), while argon and other trace gases make up the remaining one percent (see Table 3.1). Because the proportion of the most abundant gases does not change much with time and altitude (up to about 80 km), we call them **permanent gases**. These permanent gases, however, coexist with many variable constituents that, in spite of their small concentration, can greatly impact weather and climate.

Water vapor, for example, represents anywhere from almost 0 to 4% of the atmosphere, and can condense to form liquid water (dew, cloud droplets, and rain) or solid water (frost, snow, and ice). The concentration of water vapor can change greatly in time and space. For example, it is typically greater over the ocean and less over deserts, and it fluctuates at a given location as relatively drier and moister air masses move around with weather systems. Also, the water vapor concentration is typically much greater in the lower atmosphere, closer to the surface, since the ocean is the primary source of water for the atmosphere and, as we will see, warmer air near the surface can contain more water vapor than cold air.

Carbon dioxide (CO_2) represents about 0.04% of the atmosphere, and has been increasing since the Industrial Revolution (a topic we will explore further in Chapter 15). Its concentration also varies in time and space. It decreases annually in northern hemisphere spring and summer, when vegetation grows and takes up CO_2 during **photosynthesis** on land areas of the northern hemisphere, and increases in fall and winter in the northern hemisphere, when vegetation dies, decays, and releases the CO_2 back into the atmosphere. It also tends to be higher around cities and industrial regions, where the combustion of fossil fuels in factories, power plants, and cars, for example, releases CO_2.

Methane (CH_4) has even smaller concentrations, but can be found in elevated concentration near sources such as natural gas reserves, swamps, rice paddies, garbage dumps, and livestock.

Ozone (O_3) is found in very small concentrations in the lower atmosphere, which is a good thing for us, as ozone is a dangerous pollutant that can cause severe respiratory problems. Ozone, however, plays an important role higher in the atmosphere, at about 20 to 50 km altitude, in a region called the stratosphere (see Section 3.5).

The atmosphere also contains many other gases, such as **nitrous oxide** (N_2O), released mostly by bacteria in nature and by fertilized soils and various industrial processes. It also contains small solid particles called **aerosols**, mostly blown from the surface into the air, such as soil particles, dust, pollen, smoke from forest fires, and pollution particles from factories and vehicle exhausts. Salt particles contained in water droplets blown by the wind from the surface of the ocean as sea spray can also turn into aerosols, as the airborne salty water droplets evaporate, leaving behind tiny salt particles. We will see in Chapter 6 that aerosols play a crucial role in the formation of cloud droplets and, ultimately, rain.

3.3 Origin and Evolution

When was the atmosphere created? Has it always had the same composition? Should we expect to see it change further? To answer these questions, we need to step back and gain some perspective on the formation and evolution of Earth.

Table 3.1. *Composition of the atmosphere*

Permanent gases (% *of volume of dry air*)		Variable gases (% *of volume of moist air*)	
Nitrogen	78	Water vapor	0–4
Oxygen	21	Carbon dioxide	0.04
Argon	0.9	Methane	0.0002
Neon	0.002	Nitrous oxide	0.00003
Helium	0.0005	Ozone	0.000004

Strictly speaking, our short history of the atmosphere starts 13.7 billion years ago, when the universe appeared, and with it the first elements, hydrogen (H) and helium (He). About 4 to 5 billion years ago the solar system formed when the solar nebula flattened into a spinning disk and dust grains started colliding and sticking together to form clumps of matter – a process known as accretion. These clumps of matter further grew over 100 million years to become the planets of our solar system, traveling around the sun in almost circular orbits.

It is reasonable to think that, when Earth was formed, its atmosphere must have consisted of the prevalent gases in the solar system, in particular such light gases as hydrogen, helium, methane, ammonia, and water vapor. However, a large fraction of these light gases probably escaped to space. Very early in Earth's development, gravity was much lower than today, and the temperature much higher, so that it was easier for light molecules like hydrogen and helium to reach their escape velocity and overcome the weak pull of gravity. The collision of Earth with other large bodies traveling the solar system might also have helped eject the lighter gases out of the early atmosphere of Earth.

The scientific consensus is that a secondary atmosphere formed from gases released from Earth's interior by volcanic eruptions. This outgassing process enriched the atmosphere in water vapor, methane, and carbon dioxide, as well as nitrogen and ammonia. Molecular nitrogen is a stable, nearly inert gas and the non-reactive property of nitrogen explains how it simply accumulated over time and why it now represents a large fraction of the air we breathe.

It is also very likely that a large number of comets containing water in the form of ice bombarded Earth and enriched our atmosphere with water vapor. This led to an early atmosphere containing mostly water vapor (about 80%), carbon dioxide (10%), 1 to 5% of nitrogen, and about 5% other gases. Our present atmosphere, however, contains mainly nitrogen (78%) and oxygen (21%), with little water vapor, but a vast ocean underneath. What is the origin of the oceans and the water that makes clouds, rain, snow, and the rivers and aquifers that are so crucial to life on Earth?

It was possible for water to exist in the vapor phase because the early atmosphere was also very hot.

As Earth slowly cooled over millions of years, water vapor condensed into liquid water (as we will see in Chapters 5 and 6) and precipitated to form rivers, lakes, and oceans. In the process of creating reservoirs of liquid water at Earth's surface, carbon dioxide was also removed from the atmosphere. In the same way that carbon dioxide can be dissolved in a soft drink, early atmospheric carbon dioxide dissolved and accumulated in the oceans. Most of the carbon dioxide did not remain in the ocean, however, but combined with other elements to form carbonates that eventually settled as sediments in the sea floor. On very long timescales, these sediments move under the action of plate tectonics and subduct under continental margins. The sediments melt in Earth's crust and volcanic eruptions that tap into these regions restore carbon dioxide to the atmosphere.

Another important influence on atmospheric composition comes from chemical weathering of rocks in contact with air and water. For example, during oxidation, iron in rocks reacts with oxygen to form rust, which reduces atmospheric oxygen. Another example is carbonation, when carbon dioxide dissolved in rainwater reacts with limestone, which reduces atmospheric carbon dioxide. There are other chemical, biological, and physical controls on atmospheric composition, but the processes described here are thought to be the most important on long timescales.

Our story so far accounts for the small concentration of lighter gases such as hydrogen and helium, for the dominance of nitrogen, the existence of water oceans, and for the presence of carbon dioxide in sediments and in our atmosphere. But it does not account for the presence of oxygen, without which animal and human life would be impossible. About 2.3 billion years ago, primitive, anaerobic bacteria developed and started removing carbon dioxide from the atmosphere while replacing it with oxygen. The increase in oxygen then allowed the evolution of aerobic organisms and subsequent forms of multicellular life (see Box 3.1).

Our story is still missing one aspect. The incoming solar radiation was too destructive for life to survive in direct contact with sunlight. As we will discuss in Chapter 4, sunlight is composed of ultraviolet (UV) radiation, a high energy form of light that damages

Box 3.1. Photosynthesis and the Carbon Cycle

Carbon dioxide can be removed from the atmosphere by photosynthesis, by which CO_2 combines with water (H_2O) to form organic matter (CH_2O) and release oxygen (O_2):

$$CO_2 + H_2O \xrightarrow{\text{light}} CH_2O + O_2$$

Conversely, through combustion or decay (i.e., oxidation), organic matter turns into water vapor and carbon dioxide:

$$CH_2O + O_2 \longrightarrow H_2O + CO_2$$

The two chemical reactions form a short cycle through which carbon is exchanged between plant matter and the atmosphere. This short cycle essentially follows the Northern Hemisphere mid-latitude seasons and therefore its timescale is on the order of a year. By contrast, the removal of carbon by absorption of carbon dioxide in the ocean, formation of carbonates (sea shells) and sedimentation in the sea floor occurs on timescales of hundreds to millions of years, similarly to its counterpart, the release of carbon dioxide during deep sea volcanic eruptions and the collision of tectonic plates. Thus, it is important to note that a disruption of the short cycle (for example, increasing the concentration of carbon dioxide by increasing the burning of fossil fuels) cannot be fully accommodated by the long cycle (removal by absorption in the ocean), because the latter takes too long.

DNA and protein, causing, for example, sun burns and skin cancer in humans. Land areas on Earth only became habitable when the atmosphere contained enough oxygen that significant amounts of stratospheric ozone (O_3) formed, which is highly efficient at absorbing ultraviolet radiation (see Box 3.2). This speaks to the importance of the ozone layer – see Section 3.4 – and demonstrates how the existence of life on Earth is the result of a subtle balance of atmospheric composition, radiation, and heat, reached over the course of billions of years.

3.4 Future Evolution

Is the composition of our atmosphere still changing? And if so, how fast? We have been monitoring the concentration of atmospheric gases for more than 50 years now, and climate scientists have been reconstructing past concentrations from a variety of sources, such as gas bubbles trapped in Antarctic ice. We can now ascertain that, while the concentration of permanent gases such as oxygen and nitrogen has not changed much recently, the concentrations of certain variable gases, such as carbon dioxide and methane,

have changed drastically over the past 200 years. The concentration of carbon dioxide, for example, has increased from about 280 ppm (parts per million, or 0.00028) to 370 ppm (see Figure 3.1) between about 1750 and 2000 – and to over 400 ppm since then. We know that this increase is largely due to the burning of fossil fuels since the Industrial Revolution, and that this influx of carbon dioxide is taking place at a much faster rate than the removal of carbon dioxide from the atmosphere by the ocean. Since carbon dioxide, among other gases whose concentrations are also changing, absorbs infrared radiation and heats Earth (through the **greenhouse effect**, to be discussed in Chapter 4), the increase in the concentration of these greenhouse gases is also altering the heat budget of Earth and is increasing its temperature, leading to **global warming** – a subject to which we will return in Chapter 15.

In the twentieth century, human activities have also altered the concentration of ozone in the atmosphere. Recall that ozone is found in very small concentrations in the lower atmosphere, but plays a significant role in the stratosphere, at about 20 to 50 km altitude. There it absorbs UV radiation from the sun (see Box 3.2) and prevents much of it from reaching the surface, where

Box 3.2. Chapman Reactions

Ozone (O_3) is naturally created in the stratosphere by the following reactions:

$$O_2 \xrightarrow{UV} O + O \qquad (1)$$

$$O + O_2 + M \longrightarrow O_3 + M \qquad (2)$$

In reaction (1), an oxygen molecule is photolyzed by UV radiation to form two oxygen radicals. In reaction (2), one of the radicals combines with a new oxygen molecule to form an ozone molecule. This second reaction releases heat, which is absorbed by the third molecule, represented by M, which can be anything, but most likely nitrogen.

Ozone is also naturally destroyed by either of the following reactions:

$$O_3 \xrightarrow{UV} O + O_2 \qquad (3)$$

$$O_3 + O \longrightarrow O_2 + O_2 \qquad (4)$$

In reaction (3), an ozone molecule is destroyed by UV radiation to form an oxygen radical and an oxygen molecule. In reaction (4), an ozone molecule combines with a radical to form two oxygen molecules. These four reactions are referred to as "Chapman reactions" after Sydney Chapman, who first proposed them in 1930 to explain the presence of the ozone layer in the stratosphere. They are simultaneously taking place, with ozone molecules constantly created and destroyed, keeping the system in dynamic equilibrium (see Appendix 3.1).

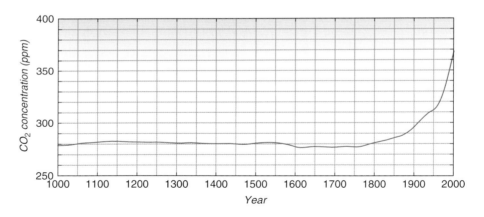

Figure 3.1. Carbondioxide concentration during the last millennium.

it would damage life (plants and animals) and cause severe problems for us, including skin cancer and eye cataracts. While the ozone concentration remains low in the troposphere, it increases in the stratosphere in proportion to the other gases. We refer to that region as the **ozone layer**. The existence of that layer was crucial for the appearance and development of life on Earth (especially on land).

The concentration of stratospheric ozone was reduced globally in the late twentieth century,

especially over Antarctica in late winter and early spring, creating the so-called ozone hole (Figure 3.2), and to a lesser extent over the Arctic. The reduction in ozone concentration was found to be due to the emission of chlorofluorocarbons (CFCs) into the atmosphere. CFCs are synthesized gases containing chlorine that were used as refrigerants, propellants in spray cans, solvents, and as blowing gas to make foam. CFCs are very stable, which is attractive for industrial use, but also means that, under the right conditions,

CHAPTER 4

Heat and Energy Transfer

Energy enters the Earth system in the form of solar radiation, preferentially heating the ground and the tropical latitudes. The resulting temperature contrasts set the atmosphere in motion, as heat is transferred upward and poleward. Weather is largely the result of this transfer of heat by atmospheric motions. In this chapter, we will explore the nature of heat and radiation, the origin of temperature contrasts on Earth, the mechanisms by which energy is transferred, and the implications for weather.

Although we might tend to think about weather in terms of wind and rain, weather is in fact largely the result of the redistribution of heat in the atmosphere. A heat imbalance, i.e., a disequilibrium in the distribution of heat between different regions of Earth, causes a transfer and redistribution of heat in the atmosphere. The resulting movement of air masses creates wind, storms, clouds, and rain. Therefore, if we are to understand the development of weather systems, we need first to understand the source of the heat imbalance and the nature of heat transfer in the atmosphere.

It is accepted in thermodynamics (the field of science that is concerned with the energy of systems) that heat does not normally flow from cold to warm objects (unless some other process is at work): it necessarily flows from warm to cold, i.e., in a direction that will bring the system to a state of equilibrium – technically, a state of lower potential energy. In the atmosphere, this transfer of heat can be achieved by three processes, conduction, convection, and radiation, which we will now explore in more detail.

4.1 Conduction

Earlier we defined heat as the kinetic energy of atoms and molecules, or their energy of motion. In solid matter, motion is reduced to the vibrations of the atoms and molecules. In a liquid or a gas, molecules

cold spoon

Heat is conducted up the spoon handle by transfer of molecular vibrations.

hot tea

Figure 4.1. Conduction of heat through a solid.

can move around and their kinetic energy describes both their vibrations and their displacement. When very energetic molecules (i.e., a hot substance, such as hot tea) are in contact with slower molecules (e.g., a cold spoon placed in the tea), the more energetic molecules necessarily impart some of their energy to the slower molecules when they bump into them (Figure 4.1). In the process, the slower molecules gain energy (i.e., the spoon warms up) while the faster molecules lose energy (i.e., the tea cools down). As a result, energy is redistributed more uniformly throughout the system. Following the same reasoning, the atoms in the bottom part of the spoon, being now more energetic than those in the top part of the spoon, will impart kinetic energy to the slower atoms by collision, in the form of faster vibrations, and heat will rise up the spoon. Note that the atoms in the spoon do not need to be displaced to transfer their heat up the spoon: they merely need to bump into adjacent atoms and share their extra energy of motion by contact. We refer to this transfer of heat by contact, collision, and transfer of molecular vibrations as **conduction**. By its very nature, conduction always transfers heat from warm to cold regions. Thus, when you touch a cold object, even though you might be under the impression that "coldness" is conducted into your hand, it is really your own heat that is conducted into the cold object.

While conduction is very efficient in solids (in particular in metals), that is not the case for fluids such as liquids and gases. In particular, air is a very poor

conductor, which is one reason why conduction is not the main process by which heat is transferred in the atmosphere. Indeed, conduction plays a significant role only at the surface, where heat is exchanged by contact between solid material and the very first layer of air in contact with it, at the centimeter scale. That is the case, for example, when heat is transferred from a hot surface to the air on a hot summer afternoon, making the lowest few centimeters of the atmosphere very warm, and when heat is transferred from the air to a cold surface in winter, making the lowest few centimeters of the atmosphere very cold. If conduction were the only process at play, however, the temperature of the troposphere would be fairly uniform, while large temperature changes would be observed in the first centimeters of the atmosphere close to the surface. When air near the surface warms by conduction, however, other processes come into play.

4.2 Convection

When an air parcel warms, its temperature, pressure, and density all change according to the ideal gas law (recall Box 1.3). The pressure inside the air parcel, however, always instantaneously adjusts to the pressure of the environment. If the air parcel is heated by conduction from a hot surface, for example, and the environmental pressure is 1013 hPa, the pressure of the air parcel remains equal to 1013 hPa (as long as it stays at sea level). Therefore, as described by the ideal gas law ($pV = nR^*T$), the volume and the temperature of the air parcel must change in proportion to each other; i.e., since the pressure is fixed, if the temperature increases, then the volume must be increasing as well. Indeed, since the mass of the parcel is unchanging (n is fixed), the heated parcel *expands*: it is subject to **thermal expansion**.

Using the other form of the ideal gas law ($p = \rho RT$), we can also say that, since the pressure is fixed, the density and the temperature of the air parcel are *inversely* proportional, i.e., if the temperature increases, then the density must decrease. Indeed, when an air parcel is heated and expands, the same mass of air is contained in a larger volume, and the density decreases.

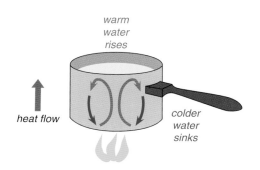

Figure 4.2. Convection and transfer of heat in a pan of water heated on a stove.

Being less dense than its surroundings, by Archimedes' principle, the air parcel is buoyant and rises. Hence the commonly used expression, "warm air rises". (We will discuss Archimedes' principle and buoyancy in more detail in Chapter 6, Box 6.4.)

We refer to this vertical fluid motion (gas or liquid) due to buoyancy as **convection** (not to be confused with "advection," which refers to horizontal motion). It is important to note that, as the warm fluid rises, not only the fluid molecules, but *also the heat they carry*, are displaced upward. Therefore, convection is primarily a heat transfer process. Moreover, because entire blobs are displaced, the corresponding heat is transferred much more efficiently than it would be by conduction (in which the molecules vibrate but are *not* displaced).

You do not need to study the atmosphere to observe convection. In fact, convection is taking place every day in your home as you use radiators, stoves, and other heating devices. When you heat a pot of water, heat is first conducted from the flames to the water through the metallic bottom of the pan, but as soon as the first millimeters of water at the bottom of the pan become warmer than the rest of the pan, warm buoyant blobs of water start rising while colder water sinks to replace them (Figure 4.2). In the process, heat is transferred and redistributed throughout the pan. The convective process is very quick and efficient. In fact, if you were to place two thermometers in the water, one at the top and one at the bottom, you would observe similar temperatures, as convective overturning is almost instantaneous. Through convective motion, the system is

responding to the heat imbalance and adjusts to a new equilibrium. It is possible because fluids move freely and can become organized as currents moving heat around; this cannot happen in solids.

The same process is constantly at work in the atmosphere. On a sunny day, the sun heats the surface, which warms the lower atmosphere by conduction (Figure 4.3(a)). Warm air rises, while colder air sinks to replace it, resulting in a net upward transfer of heat (Figure 4.3(b)). We refer to these rising pockets of warm air as **thermals**. In Chapter 6, we will learn that, as warm air rises, it also expands and cools in response to lower pressure, which can lead to condensation and the formation of clouds. But before we get distracted by clouds, let us explore the third way by which heat is transferred in the atmosphere.

4.3 Radiation

All objects in the universe contain some level of heat, even very cold objects. The vibration of molecules creates an electromagnetic field and the emission of **electromagnetic radiation**. While this is quite difficult to conceptualize using classic imagery, we can think of radiation as waves of energy that propagate through space at a speed of 300 000 kilometers per second (i.e., 186 000 miles per second). These waves correspond to the joint oscillation of the electrical and magnetic fields, and the properties of the waves depend on the temperature of the object creating them. Colder objects generate low-energy waves, while warmer objects generate high-energy waves. In fact, you know at least one family of electromagnetic waves: the sun, at a temperature of 6000 K, generates waves that peak in **visible** light, which our eyes have evolved to detect. The sun also generates higher energy radiation, such as **ultraviolet**, and lower energy radiation, such as **infrared**. Although our eyes cannot detect ultraviolet and infrared radiation, our skin is sensitive to both.

Since we have a temperature and also contain heat, our bodies also emit radiation at all times – infrared radiation. Although our eyes cannot see it, it can be measured, and can be observed to vary with changes in our temperature.

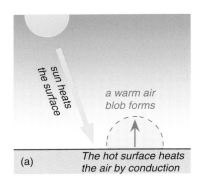

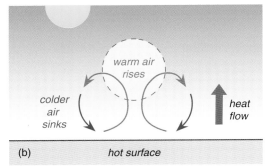

Figure 4.3. Formation of a thermal by absorption of radiation and conduction (a) and convection (b).

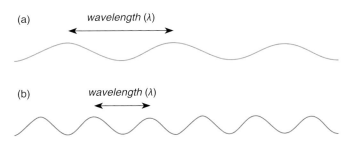

Figure 4.4. Schematic representation of two electromagnetic waves of different wavelength (λ). (a) $\lambda = 2.5$ cm; (b) $\lambda = 1.25$ cm.

Table 4.1. *Spectrum of electromagnetic radiation*

Radiation band	Wavelength
Gamma rays	<0.0001 µm
X-rays	0.0001 to 0.01 µm
Ultraviolet	0.01 to 0.4 µm
Visible	0.4 to 0.7 µm
Near infrared	0.7 to 4.0 µm
Thermal infrared	4 to 100 µm
Microwave	100 µm to 1 m
Radio waves	>1 m

4.3.1 The Nature of Electromagnetic Radiation

Among the properties of electromagnetic radiation, two important measures are the wave **frequency** and **wavelength**, which are related by the wave speed. These waves can travel through space, and in particular they can travel through a complete vacuum, where they travel at a speed of 300 000 km/s, i.e., the speed of light. We can think of these waves as undulating in every direction of space around the source, the troughs and crests of the waves being the signature of the pulsations at a certain frequency. For example, Figure 4.4 represents a wave with a crest-to-crest distance of about 2.5 cm.

This crest-to-crest distance is called the wavelength and is commonly represented by the Greek letter lambda (λ). Low energy radiation has wavelengths on the order of a meter or more, such as radio and TV waves. High energy radiation, however, pulsates much faster, which means that the crests and troughs happen more frequently (hence, "higher frequency") and the wavelength is much shorter. In Figure 4.4(b), for example, the wavelength is about 1.25 cm, half the wavelength of the wave in Figure 4.4(a). Since the blue wave travels at the same speed as the red one (i.e., at the speed of light), the frequency is twice as large.

Since high energy radiation has wavelengths on the order of a thousandth, millionth, or billionth of a meter, we resort to submultiple units:

1 meter = 1000 millimeters (mm)
= 1 000 000 micrometers (µm)
= 1 000 000 000 nanometers (nm)

In this unit system, visible light has a wavelength on the order of 0.5 µm, or 500 nm, while infrared radiation has a wavelength on the order of 10 µm (see Table 4.1).

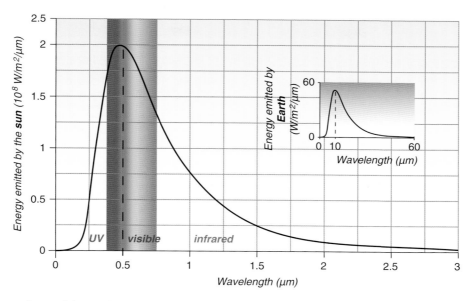

Figure 4.5. Comparison of the emission spectra of the sun and Earth.

4.3.2 Temperature and radiation

The radiation emitted by an object is a function of its temperature. As a rule of thumb, warmer objects emit *more* electromagnetic radiation (i.e., a greater quantity of energy), and they do so *at higher frequency*. This behavior is captured by Planck's law, named after the famous German physicist, Max Planck (1858–1947). Figure 4.5 shows an example of Planck's law applied to the sun. It shows the amount of energy emitted by an object with the temperature of the sun (6000 K), as a function of wavelength. The sun emits a large quantity of visible light. It also emits ultraviolet radiation, and a fair amount of infrared radiation (as it turns out, about 50% of the radiation emitted by the sun is actually in the infrared).

Although the amplitude of the emission spectrum varies with the temperature of the object, its shape is similar at all temperatures: it has a peak of maximum emission (at 0.5 μm in the case of the sun) and two tails showing the decreasing amounts of radiation emitted at lower and higher frequencies. (See Box 4.1 for a description of other formulas derived from Planck's law and describing the emission spectrum.) Thus, Earth's emission spectrum is very similar in shape to that of the sun, but it is very different both in magnitude and

spread (see insert in Figure 4.5). Because the average temperature of Earth is 288 K, the peak of maximum emission is shifted to 10 μm (in the infrared), and the intensity of the emission is several orders of magnitude smaller than that of the sun. Thus, if we were to superimpose the emission spectrum of Earth on that of the sun in Figure 4.5, it would be stretched out to the right (i.e., toward longer wavelengths), in the infrared, and would be so flat (i.e., so small) that it would blend with the abscissa. This is consistent with the fact that the sun is a star, and therefore releases large quantities of energy through nuclear fusion, whereas Earth is a planet at much lower temperature, which mostly receives energy from the sun and radiates it back to space in the form of infrared radiation.

Let us now explore how this visible and infrared radiation interacts with the atmosphere.

4.4 Radiative Interactions

When radiation encounters an object, it can interact with it in different ways. It can be **transmitted** through the object, as it would be through a window, for example, or through space. Radiation can also be **absorbed**, **reflected**, and **scattered**.

Box 4.1.   Radiation Laws

Figure 4.1.1.   Gustav Kirchoff.

In the second half of the nineteenth century, a series of experimental and theoretical findings revolutionized our understanding of radiation and were critical for the advancement of physics. Around 1860, Gustav Kirchhoff introduced the concept of a "black body" – a body that, simply put, absorbs all radiation – and stated that, if we know how much this theoretical black body emits at each wavelength at a given temperature, then we know how much *any* body emits, in proportion to how much it absorbs. Thus, for example, if a body absorbs half as much as a black body, it also emits half as much as a black body. **Kirchhoff's law** is better known in the form, "good absorbers are good emitters; poor absorbers are poor emitters."

Figure 4.1.2.   Jožef Stefan.

A black body at a given temperature, however, does not emit the same amount of radiation at all wavelengths. The exact emission function would later be determined by Max Planck – see below. In the meantime, a series of scientists laid the ground for him by establishing several important relationships. At the end of the nineteenth century, Jožef Stefan (1835–1893) and Ludwig Boltzmann (1844–1906) determined that the total emission of radiation (E) by a black body at a given temperature T is proportional to the fourth power of temperature:

$$E \propto T^4$$

Figure 4.1.3.   Ludwig Boltzmann.

We refer to this relationship as the **Stefan–Boltzmann law**. It tells us that, even with a small increase in temperature (expressed in kelvin), the total amount of energy emitted by an object (in W/m^2) increases with the fourth power of temperature. For example, if the temperature is multiplied by 2, the total amount of energy is not simply multiplied by 2, but by 2^4, which is equal to 16.

This relationship has many applications in physics. In certain situations, we prefer to use the inverse relationship by which the temperature of an object changes as the fourth root of the radiative energy it emits:

$$T \propto \sqrt[4]{E}$$

Thus, if we can somehow measure or estimate the total amount of energy emitted by an object, such as with a satellite in space, we can estimate the object's temperature.

In 1893, Wilhelm Wien (1864–1928) derived another important relationship which tells us that the wavelength of peak emission (λ_m) of a black body is inversely proportional to the temperature T of that black body:

$$\lambda_m \approx 2897/T$$

where λ_m is expressed in micrometers and T in kelvin. In this relationship of inverse proportionality, known as **Wien's displacement law**, when T increases, λ_m decreases, which corresponds to a shorter wavelength, and therefore a higher frequency. The sun, for example, with a temperature of 6000 K, emits mostly around a wavelength of $2897/6000 \approx 0.5\,\mu m$ (i.e., in the visible part of the electromagnetic spectrum). By contrast,

Box 4.1. (cont.)

Figure 4.1.4. Wilhelm Wien.

Earth, with an average temperature of 288 K, emits mostly around a wavelength of 2897/288 ≈ 10 μm (i.e., in the infrared part of the spectrum).

In a different context, we sometimes prefer to use the inverse relationship:

$$T \approx 2897/\lambda_m$$

Now, if we can somehow measure the wavelength of peak emission of a body (for example, a distant star), we can estimate its temperature. These inverse relationships are used to design instruments that estimate the temperature of objects based on their radiative properties. Since they estimate temperature, these instruments are really thermometers, although they use a different technology from that of the mercury or alcohol thermometers you might have at home. The most common such instrument is the so-called "infrared thermometer," which is specifically designed to measure radiation in the infrared band of the electromagnetic spectrum, i.e., typical radiation emitted by objects on Earth.

These radiation laws are really derived forms of a more exact and complete relationship established by Max Planck (1858–1947) around 1900, and shown graphically in Figure 4.5. **Planck's law** describes the amount of radiation emitted by a black body in thermal equilibrium at a given temperature as a function of wavelength. It helps us understand how a body can be a good absorber and a good emitter (Kirchhoff's law) and yet not emit the same amount of radiation as it absorbs at a given wavelength, because the emitted amount depends only on temperature (see Box 4.3). Planck's law can be integrated to produce an exact form of the Stefan–Boltzmann law and differentiated to produce Wien's law.

Planck's findings were revolutionary in many respects and opened the door to the development of quantum field theory and modern physics.

Figure 4.1.5. Max Planck.

4.4.1 Absorption

When radiation is absorbed by an object, it interacts with the atoms and molecules of that object in such a way that it is converted to kinetic energy. Recall the law of conservation of energy (Box 1.1), by which energy cannot be created or destroyed, but only transferred or, in our present case, transformed (from radiative energy to energy of motion). As a result, the atoms and molecules either move or vibrate faster. In other words, when absorbed, radiation is converted to **heat**. We certainly know that from standing in the sun: our body warms because it absorbs the incoming sunlight.

From standing in the sun, we also know that our clothing determines how much we will warm. When we wear white clothes, for example, we stay cooler than when we wear dark clothes. This tells us that materials do not absorb radiation in the same way. Indeed, some materials only absorb part of the incoming radiation, and the rest is reflected.

4.4.2 Reflection

Whatever is not transmitted or absorbed by a material is **reflected**, as it would be by a mirror. To continue with the clothing example, we know that white clothes

Table 4.2. *Albedo of typical surface types*

Surface type	Albedo(%)
Fresh snow	80–90
Thin cloud	30
Thick cloud	90
Ice	30–60
Grass	10–30
Water	10
Forest	10–20
Venus	78
Earth as a whole	~30

reflect most of the visible radiation that strikes them, which keeps us relatively cool. Dark clothes, however, absorb much of the incoming visible radiation, and we quickly become warm.

Thus, absorption and reflection are intimately related. With solid, opaque surfaces, whatever is not absorbed is reflected, and vice versa. A red apple, for example, absorbs most visible radiation except red light, which is reflected back in all directions, including into our eyes: that is why we see the apple as red.

On Earth, some of the sun's radiation is reflected by bright surfaces such as clouds, snow, and ice. This is most evident in a visible satellite image such as that shown in Figure 2.13. Note that land areas are also relatively bright, compared to the ocean. Land can reflect a significant fraction of the incoming sunlight, whereas light tends to penetrate into ocean waters where it is first transmitted and then eventually absorbed.

Since features on Earth reflect various amounts of visible light, it is useful to define a new quantity, the **albedo**, as the fraction (or percentage) of incoming sunlight that is reflected by an object or surface. Table 4.2 shows typical values of albedo for common surface types. Note how forests, for example, have a low albedo (10 to 20%). Indeed, they often look very dark from a distance, or from a plane, because most of the incoming sunlight penetrates through the branches and foliage and never comes back out. Note

also how Venus has a high albedo (78%): it is surrounded by clouds that reflect a large fraction of the incoming sunlight. In contrast, the average albedo of Earth is relatively low (~30%), which means that most of the incoming sunlight is not reflected back out to space, but is absorbed by Earth.

4.4.3 Scattering

When radiation encounters individual atoms, molecules, aerosols, and cloud droplets, it is not exactly reflected as it would be by a uniform, flat surface, but is deflected in various directions, or **scattered**. Scattering is a result of complex interactions between the small particles and the electromagnetic waves. Radiation can be scattered forward, backward, and to the sides, depending on the wavelength of the radiation and the size of the particle. Cloud droplets are large enough to scatter effectively all wavelengths equally, which is why clouds appear white. Molecules and aerosols, however, preferentially scatter certain wavelengths, which explains why the sky is blue and sunsets are red (see Box 4.2).

4.4.4 Radiative Equilibrium

Transmission, absorption, reflection, and scattering also relate to emission of radiation and the balance between incoming and outgoing radiation. Recall that all objects emit some form of radiation. By inverse reasoning, since absorption results in a gain of heat, you should not be surprised that emission results in a *loss* of heat. Since objects emit radiation all the time, they are constantly losing energy. You certainly know this from sleeping without a blanket. In addition to losing heat by conduction, you also lose heat by emission of infrared radiation, and you can get cold very fast – which is why you use a blanket to keep warm. The same reasoning applies to Earth. It is constantly losing energy to space in the form of infrared radiation. Why is Earth not much colder than it is? Because half of Earth is also exposed to the sun at all times, and the planet is constantly absorbing sunlight. Thus, there is a balance between the amount of radiation absorbed

Box 4.2.   Applications of Scattering

When radiation interacts with individual atoms, molecules, aerosols, or cloud droplets, it can be scattered in various directions, depending on the wavelength of the radiation and the size of the particle. Nitrogen and oxygen molecules, for example, are very efficient at scattering visible light from the sun at shorter wavelengths around 0.45 µm, which corresponds to blue light. That is why the sky is blue: when you look up at the atmosphere, away from the sun, you see sun rays that have been deflected by nitrogen or oxygen molecules, either once or multiple times (Figure 4.2.1).

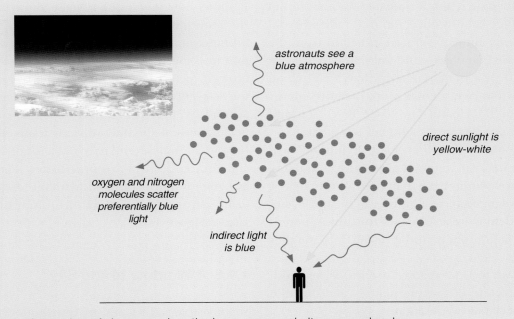

Figure 4.2.1.   Scattering of short wavelengths by oxygen and nitrogen molecules.

When the sun is low on the horizon, at sunset or sunrise, sunlight must travel a longer distance through the atmosphere before reaching your eyes, and most of the blue light has been scattered away, leaving only longer wavelengths to reach you, such as yellows and reds (Figure 4.2.2).

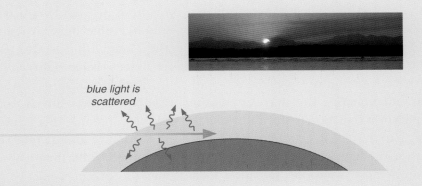

Figure 4.2.2.   Depletion of short wavelengths at sunrise and sunset.

by Earth on the day side, and the amount of infrared radiation emitted by the whole planet. Earth is in **radiative equilibrium** – another example of dynamic equilibrium (see Appendix 3.1). If Earth emitted more radiative energy than it absorbed, it would cool – and it would warm if it emitted less than it absorbed.

The fact that Earth emits nearly the same amount of energy as it absorbs is no coincidence: nature seeks equilibrium, and Earth is, by nature, in radiative balance. Our red apple is also in radiative equilibrium in the kitchen. It constantly emits infrared radiation, but it also constantly absorbs infrared radiation from the surrounding kitchen objects and walls. If we place it in full sunlight, it absorbs more energy than it emits, and warms. But as it warms, it emits more radiative energy (by the Stefan–Boltzmann law – see Box 4.1), and reaches a new equilibrium at a higher temperature.

We will see shortly that, when some regions of Earth absorb more energy than they emit, heat accumulates, and if other areas emit more than they absorb, heat is lost. This imbalance causes a transfer of heat from surplus areas to deficit areas, which in the atmosphere is achieved through motion, and gives birth to weather systems.

But before we turn to weather systems, let us finish our radiative picture of the atmosphere, and mention one important climate application.

4.4.5 Selective Absorbers

So far we have assumed that Earth absorbs and emits radiation at maximum efficiency. An object that has such characteristics is called a **black body**, or **ideal radiator** (see Box 4.1). While this is a useful concept in theory, many objects are in fact not fully efficient (see Box 4.3). In particular, the atmosphere does not behave like a black body, because it is composed of different gases that interact in very specific ways with electromagnetic radiation (see Box 4.4). Ozone, for example, mostly absorbs radiation in the ultraviolet, at wavelengths less than $0.3\,\mu m$. Recall that this absorption was crucial for the appearance of life on Earth, because it prevents most ultraviolet radiation from reaching the surface, where it is detrimental to plant and animal life. Ozone also absorbs some infrared

radiation in a very narrow band at about $9\,\mu m$, which makes it a greenhouse gas. But it is largely transparent to other types of radiation, such as visible light, for example, and most of the infrared spectrum. Water vapor, in contrast, absorbs a significant fraction of infrared radiation, but is mostly transparent to visible and ultraviolet radiation.

Figure 4.6 illustrates the impact of the absorption of radiation by atmospheric gases on solar radiation. It shows how much radiation is transmitted through the atmosphere at various wavelengths. A low transmittance at a certain wavelength is due to high absorption by a specific gas. The main selective absorbers of interest to us are indicated at the top of the figure for reference. Water vapor absorbs in a broad band around 6–$7\,\mu m$, in addition to several other isolated peaks. Carbon dioxide also efficiently absorbs infrared radiation in several narrow bands of the spectrum, and in a large band around $15\,\mu m$. The absorption by ozone is more confined to about 5 and $10\,\mu m$, but ozone strongly absorbs in the ultraviolet (leftmost part of Figure 4.6).

4.4.6 A Window to the Sky

Since visible light is mostly transmitted through the atmosphere, we can use the analogy of a window and say that there is a "visible window" in the absorption spectrum of the atmosphere, around $0.5\,\mu m$. Similarly, because some infrared radiation is transmitted through the atmosphere in a broad band around $10\,\mu m$, without much absorption, we can say that there exists an **infrared window**. The presence of these two windows is very important, as the first allows visible light to reach the surface and warm the planet (that is also why our eyes have evolved to detect visible light) and the second allows some infrared radiation to escape to space (recall from Chapter 2 that this is useful for infrared satellite observations). If this infrared window did not exist, the radiative energy received from the sun would be reemitted by Earth and absorbed by the atmosphere, and the planet would be warmer than it is. These two windows allow Earth to be in radiative balance in a temperature range that permits water to exist

Box 4.3. Radiation and Temperature

The radiation laws described in Box 4.1 are useful in understanding many radiative processes happening on Earth. A good application of Kirchhoff's law, for example, is the radiative properties of a space blanket (often referred to as a Mylar® blanket). As you might know, such a blanket is designed to keep you warm, in particular by limiting the loss of infrared radiation from your body. Indeed, the blanket is made of a material that is a good reflector, and therefore a poor absorber. According to Kirchhoff's law, a poor absorber is also a poor emitter. Therefore, the blanket emits very little infrared radiation, while your own infrared radiation is reflected back to your body.

Kirchhoff's law can be confusing, however, when dealing with bodies at different temperatures. In this case, a body can be a good absorber and therefore a good emitter, and yet not emit the same amount of radiation as the amount it absorbed at a given wavelength. That is because, according to Planck's law, the amount of emitted radiation depends on the temperature of the body. Let us illustrate this subtlety with an example of great importance in climate science.

Let us assume that Earth is a black body with zero albedo. One could ask, since Earth absorbs mostly visible radiation from the sun, why Earth does not also emit a lot of visible radiation. According to Kirchhoff's law, being a good absorber of visible radiation, Earth should also be a good emitter of visible radiation. And indeed it is, but it does not have the right temperature to emit a lot. Planck's law tells us that, to emit a significant amount of visible radiation, a body needs to have a temperature approaching that of the sun, i.e., 6000 K. The temperature of Earth, however, is closer to 288 K, much less than that of the sun. Therefore, it emits

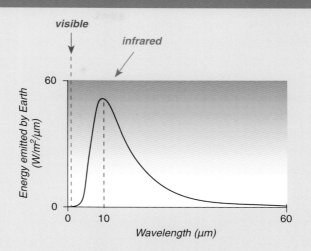

Figure 4.3.1. Emission spectrum of Earth.

mostly infrared radiation. Note that, strictly speaking, Earth *does* emit visible radiation at maximum efficiency, or has the potential to do so, but that maximum is so small that Earth emits virtually none. To illustrate this point graphically, we could return to the insert of Figure 4.5 (repeated in Figure 4.3.1 for convenience), showing the emission curve (i.e., Planck's law) for Earth. The temperature of Earth is such that its peak emission is in the infrared (around 10 μm). Note that visible radiation (around 0.5 μm) is in the leftmost tail of the curve, and is virtually zero.

Finally, note that most objects are not black bodies, i.e., they do not emit at maximum efficiency at all wavelengths, as a black body would. In other words, there exist specific wavelengths at which they absorb and emit well, and others at which they don't. The atmosphere, being made of different gases, is a good example, which creates the possibility of visible and infrared windows, as well as the greenhouse effect.

in all three phases, which has been conducive to the evolution of life.

Although some infrared radiation escapes to space, a significant fraction is absorbed by selectively absorbing gases, which gives rise to the greenhouse effect.

4.4.7 The Greenhouse Effect

Let us first imagine a planet Earth *without* an atmosphere. If there were no gases either to absorb or emit radiation, Earth would simply absorb radiation

Box 4.4. Selective Absorption and Emission of Radiation

There exist two mechanisms by which gases interact with electromagnetic waves. The first mechanism involves changes in the energy state of electrons at the atomic level and explains, for example, the absorption of ultraviolet radiation by ozone. The second mechanism involves changes in the vibration and rotation of the molecules and explains the absorption of infrared radiation by greenhouse gases.

Electronic Transition

You might know that atoms are made of a nucleus and a specific number of electrons orbiting the nucleus at a particular distance from it. We can

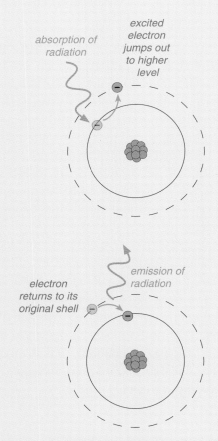

absorption of radiation

excited electron jumps out to higher level

electron returns to its original shell

emission of radiation

Figure 4.4.1. Electronic transition during the absorption and emission of radiation.

picture them as existing on a sphere around the nucleus, that sphere having a specific radius. We refer to such a sphere as a "shell." Moreover, electrons can exist on separate shells, each shell being successively farther away from the nucleus. We find two electrons on the first shell, then eight on the second shell, and so on. Once a shell is full, electrons populate the next one. The last shell might be full or not. The total number of electrons determines the type of atom we are dealing with, i.e., the element, hydrogen, helium, carbon, oxygen, etc.

To interact with an electromagnetic wave by electronic transition, an atom or molecule must contain an electron that can use the energy carried by the wave to leap from its current level (i.e., its current shell) to a higher level – to another shell that is more distant from the nucleus (see Figure 4.4.1). Such a leap requires energy, and that energy is captured from the electromagnetic wave. However, because the successive shells occur at very specific distances from the nucleus, these leaps require a very specific amount of energy (we say that they are *quantized*). These leaps can only happen if the electromagnetic wave carries exactly the appropriate amount of energy. In other words, that electromagnetic wave must have a very specific *wavelength*. In the case of ozone, for example, the electrons will only be able to perform the leap if the ozone molecules are hit by ultraviolet radiation.

Vibration and Rotation

Certain molecules can also absorb electromagnetic radiation to activate specific modes of vibration and rotation called "internal modes." The atoms constituting a water vapor molecule, for example, represented in Figure 4.4.2 schematically by two smaller white circles (hydrogen atoms) and one larger red circle (oxygen atom), can vibrate with respect to each other in different ways.

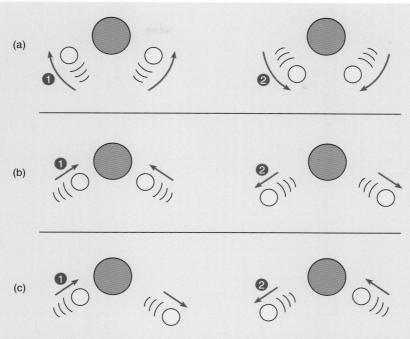

Figure 4.4.2. Internal modes of vibration and rotation of the water vapor molecule. (a) Atoms bending up and down. (b) Atoms vibrating symmetrically. (c) Atoms vibrating asymmetrically. White circles represent hydrogen atoms and red circles represent oxygen atoms. Numbers 1 and 2 indicate the two steps of each vibration mode.

The molecule can also spin on itself, and all these different modes of vibration and rotation require a specific amount of energy. Here again, the water vapor molecule will only absorb electromagnetic waves that carry the exact amount of energy required to excite these internal modes, i.e., electromagnetic radiation with the adequate wavelength. Water vapor and the other greenhouse gases absorb primarily in the infrared, which explains their role in the greenhouse effect and global warming.

The wavelengths and intensity of absorption of individual gases can all be determined precisely in a laboratory with an instrument called a *spectrophotometer*. Because all these gases absorb radiation selectively at certain wavelengths but not others, we call them **selective absorbers**.

received directly from the sun and would emit infrared radiation back out to space (Figure. 4.7(a)). We can calculate the amount of solar radiation intercepted by Earth, and, considering the planet is in radiative equilibrium, we may use the Stefan–Boltzmann law to estimate the corresponding temperature (see Box 4.1). This is a classic calculation in climate science and yields an equilibrium temperature of about −18°C – extremely cold compared to the 15°C in reality.

Let us now add an atmosphere to our planet Earth, made of nitrogen, oxygen, and water vapor. Nitrogen and oxygen do not change the picture much, as they do not absorb radiation in the visible and infrared. The fact that water vapor is a good absorber of infrared radiation, however, has important implications. Since Earth's surface (land, ocean, vegetation, lakes, ice, snow, etc.) constantly emits infrared radiation, some of this radiation is absorbed by water vapor in

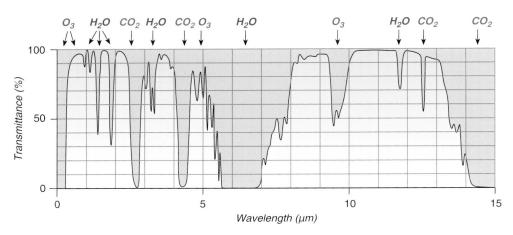

Figure 4.6. Infrared transmission spectrum of the atmosphere. Some of the wavelengths at which the main selective absorbers (water vapor, carbon dioxide, and ozone) absorb radiation are indicated at the top.

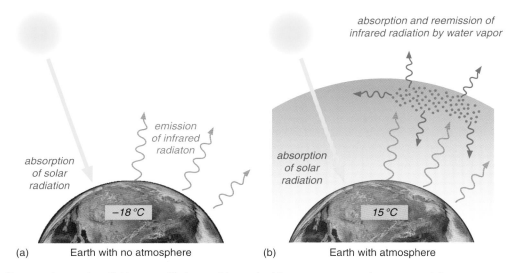

Figure 4.7. Comparison of radiative equilibrium with and without an atmosphere containing water vapor.

the atmosphere. By absorbing infrared radiation, water vapor heats the atmosphere (the heat is distributed by conduction, convection, and radiation). Thus, the presence of water vapor implies a warmer atmosphere compared to one lacking water vapor. Our story does not stop there, though. Good absorbers are also good emitters, so water vapor also strongly emits infrared radiation. Some of this radiation eventually makes its way to space. But some of it is emitted downward, where it is reabsorbed by the surface, and to the sides, in every direction, where it can be reabsorbed by other water vapor molecules. Since the surface absorbs infrared radiation, it warms, and therefore emits more infrared radiation, which is then reabsorbed by water vapor, reemitted, and so on, until a new equilibrium is reached (Figure. 4.7(b)). Under this new equilibrium, the lower atmosphere and the surface are much warmer than they would be without water vapor.

Because there is an apparent similarity with a greenhouse, in which sunlight enters but heat does not leave easily, resulting in a higher temperature,

Box 4.5.   Feedback Loops

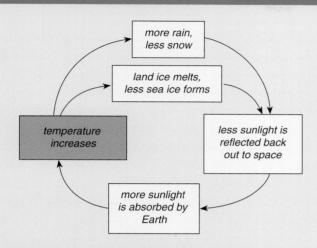

Figure 4.5.1.   Ice-albedo feedback (example of a positive feedback loop).

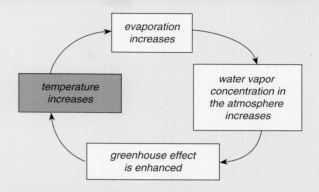

Figure 4.5.2.   Water vapor feedback.

The enhancement of the greenhouse effect is further complicated by other processes that are altered as a result of increasing temperatures. In a warmer world, ice and snow start to melt, for example. You might recall that ice and snow have a high albedo: they are excellent reflectors of visible light. If the ice and snow coverage decreases, less sunlight will be reflected, and therefore more sunlight will be absorbed by Earth's surface. An increase in the absorption of sunlight translates into higher temperatures, and therefore even more warming. This whole chain of events is called a **positive feedback loop** (Figure 4.5.1) and accelerates global warming in ways that are not always obvious to model and understand.

In addition to the **ice-albedo feedback** just described, we expect to observe more evaporation from the ocean in a warmer world, and therefore more water vapor in the atmosphere. Water vapor being a greenhouse gas, this will lead to even more absorption of infrared radiation, and therefore even more warming – another positive feedback loop. In fact, the **water vapor feedback** is the main positive climate change feedback (Figure 4.5.2).

A warmer planet, however, will also emit more infrared radiation (recall Planck's law), and will therefore lose more energy, leading to a temperature decrease. In this particular case, warming is not reinforced by more warming, and the **Planck feedback** is a **negative feedback** (Figure 4.5.3).

The Earth system contains a large number of feedback loops, both positive and negative, and climate scientists spend a fair amount of time trying to model and understand which will dominate and control the magnitude and rate of global warming in decades and centuries to come.

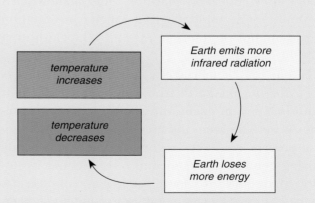

Figure 4.5.3.   Planck feedback (example of a negative feedback loop).

the radiative process we just described is called the **greenhouse effect**. Know that this is somewhat of a misnomer, however, as the reason why a greenhouse is very warm is not because infrared radiation is trapped in the greenhouse: it is primarily because the warm air itself is trapped inside, preventing convection and the mixing of warm and cold air that would naturally take place outside in the free atmosphere. We sometimes refer to this effect as "suppressed convection." Therefore, it is a convective effect, before being a radiative effect. The phenomenon that takes place in the atmosphere is primarily a radiative effect, and some have proposed to call it the "atmosphere effect." Nevertheless, the term "greenhouse effect" has caught on and is now widely used.

Due to its higher concentration in our atmosphere, water vapor is the main actor of the greenhouse effect taking place on Earth. It is not the only one, however, as many other gases also absorb infrared radiation, such as carbon dioxide, methane, ozone, and nitrous oxide. We call these **greenhouse gases**. In the presence of all greenhouse gases combined, the equilibrium temperature reached by the lower atmosphere is about 15 °C, or 33 °C higher than it would be without an atmosphere.

To summarize, greenhouse gases in the atmosphere absorb energy in the form of infrared radiation and maintain Earth's surface at a higher temperature than would be the case without an atmosphere. The greenhouse effect also takes place on other planets, such as Venus, for example, where the high concentration of carbon dioxide results in extreme temperatures of 750 °C.

Given how beneficial the greenhouse effect has been for life on Earth, you may wonder why it is expressed as a problem with respect to global warming. As we have seen, the greenhouse effect is a natural process by which Earth has maintained a slowly varying equilibrium temperature throughout its history, in response to slowly varying greenhouse gas concentrations. The temperature has sometimes been higher, when greenhouse gas concentrations were higher (such as in the Eocene), and sometimes lower, when greenhouse gas concentrations were lower (such as in the Ice Ages). More recently, human civilization has altered this slowly changing natural process

by *increasing* the concentration of greenhouse gases in the atmosphere very rapidly, most certainly over the last two centuries, since the Industrial Revolution, and possibly for the last 8000 years, with deforestation and the development of agriculture. Increasing greenhouse gases means more absorption of infrared radiation, and therefore more heat and a higher temperature for Earth. This human enhancement of the greenhouse effect is called **global warming**, and it is leading to **climate change**. Furthermore, it is reinforced by positive feedback processes that may accelerate the warming, and make the planet warm much faster than the radiative effect alone would suggest (see Box 4.5 and Chapter 15).

4.5 Radiation and Weather

Let us return now to our main topic and discuss how radiative interactions control temperature, and therefore weather.

4.5.1 Heat Imbalance

We mentioned earlier the importance of radiative equilibrium, where heat inputs balance outputs: the amount of incoming solar radiation must be balanced by outgoing infrared radiation. What happens if a specific region of Earth is *not* in radiative equilibrium? What if it receives more energy than it emits, or vice versa?

In the tropics, for example, the sun is high in the sky, and the sun rays tend to be concentrated over smaller areas compared to the polar regions, where the sun is low in the sky and sun rays are spread out over larger areas (Figure 4.8). Think of two sun beams carrying the same amount of energy: since that energy is spread out over a larger area at the poles, each unit area is receiving less radiation. In contrast, since the sun beam is concentrated over a smaller area in the tropics, each unit area is receiving comparatively more energy than at the poles. This means that the absorption of solar energy, and therefore heating, is much more intense at low latitudes – in the tropics – than at high latitudes – in the polar regions.

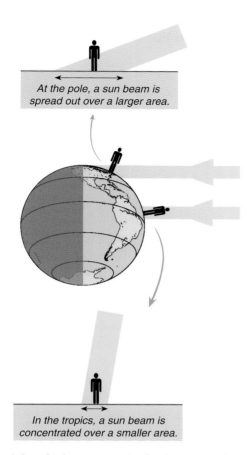

At the pole, a sun beam is spread out over a larger area.

In the tropics, a sun beam is concentrated over a smaller area.

Figure 4.8. A given amount of solar energy is concentrated into a smaller area in the tropics, whereas it is more spread out at the poles.

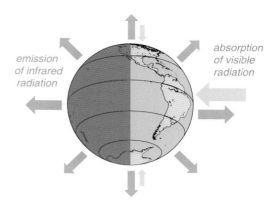

emission of infrared radiation

absorption of visible radiation

Figure 4.9. The emission of infrared radiation by Earth (in red) is relatively uniform around the globe, while the absorption of visible radiation from the sun (in yellow) is greater at low latitudes than at the poles, which results in a net input of energy at low latitudes, and a net output of energy at the poles. (Note that the yellow arrows at the poles are shown perpendicular to the surface for clarity, even though the sun rays are coming from the right of the picture.)

In contrast, cooling by emission of infrared radiation is always emitted "upward" and is not subject to the same angle variation. It is pretty similar all around Earth, except for the fact that temperatures are higher in the tropics, and infrared emission is somewhat higher there (according to the Stefan–Boltzmann law – see Box 4.1). As a result, heating by absorption of solar radiation is greater than cooling by emission of infrared radiation in the tropics, and there is **net warming**. But heating by absorption of solar radiation is *less* than cooling by emission of infrared radiation at the poles, and there is **net cooling** (Figure 4.9).

This contrast is enhanced by the fact that sun rays have to travel a longer distance through the atmosphere to reach the poles (being at a low angle) and a shorter distance in the tropics (being overhead). On its way to the surface, sunlight is scattered and much of the scattered energy is lost to space. [This is somewhat similar to the red sunset example we encountered in Box 4.2.] The longer the path, the more scattering, and the more the sun beam is depleted of its energy. Therefore, more energy is being lost at the poles by scattering, compared to the tropics, which reinforces the heat imbalance between high and low latitudes.

Finally, recall that the albedo of Earth is high at the poles, where we find much ice and snow (Figure 4.10), and low in the tropics, where we find large regions of ocean. More absorption in the tropics and more reflection at the poles also contribute to creating a heat imbalance between high and low latitudes.

As a result, heat tends to accumulate in the tropics, while there is a deficit at the poles. Since the whole system tends to return to a state of equilibrium by natural law, it will evolve in such a way as to transfer continuously the surplus of tropical heat toward the poles. This is accomplished by motion of air in the atmosphere and motion of water in the ocean, through air currents in the atmosphere and water currents in the ocean. For the atmosphere,

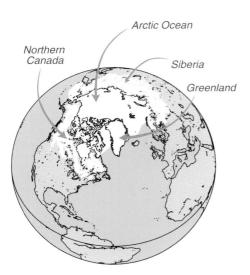

Northern Canada

Arctic Ocean

Siberia

Greenland

The North Pole is an ocean covered with ice and surrounded by land and snow in winter, whereas the South Pole is a continent covered with glaciers and snow, and surrounded by sea ice.

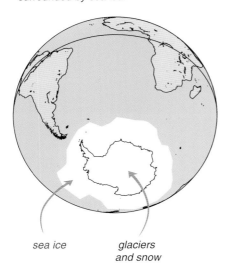

sea ice

glaciers and snow

Figure 4.10. Albedo at the poles.

a central part of this exchange of heat, executed through the exchange of warm and cold air masses, is the **midlatitude cyclone**, to which we will return in Chapter 10. In the meantime, let us remember that weather on Earth is largely due to the existence of a heat imbalance between the tropics and the poles, and that this heat imbalance can be explained by differences in the absorption and emission of radiative energy at different latitudes.

4.5.2 Seasonal Variations

The interaction of solar radiation with Earth also explains why we have **seasons**. As shown in Figure 4.11, the sun is higher in the sky in summer, and lower in winter. Therefore, for similar reasons to those invoked earlier when describing latitudinal differences, solar radiation is concentrated onto a smaller area in summer, which results in more energy being absorbed per unit area, and higher temperatures. In winter, solar radiation is spread out over a larger area, which results in less energy being absorbed per unit area, and lower temperatures.

Moreover, because Earth is tilted toward the sun in summer, a given location in the summer hemisphere spends more time on the day side than on the night side (Figure 4.12). In other words, the days are longer than the nights, which means that such a location is absorbing heat during a longer time period in summer than in winter. This also contributes to the seasonal contrast, with higher temperatures in summer and lower temperatures in winter.

In summary, seasons are due to the tilt of Earth's axis with respect to the ecliptic plane (i.e., the plane of Earth's rotation around the sun), which brings the summer hemisphere under more direct solar radiation during longer days, and the winter hemisphere under less direct radiation during shorter days. [We note that seasons are *not* due to changes in the distance between Earth and the sun during Earth's revolution, as many people believe. Currently, Earth is actually slightly farther away from the sun during northern summer.]

4.5.3 Diurnal Variations

Temperature also changes on a daily basis, which we refer to as daily temperature variations, or **diurnal** temperature variations. Obviously, the sun rises and sets, which means that Earth's surface receives more solar energy during the day than during the night. But there is more to the diurnal cycle than just the times of sunrise and sunset.

Earth also loses energy by emission of infrared radiation, which brings us back to the concept of dynamic equilibrium (Appendix 3.1). To return to

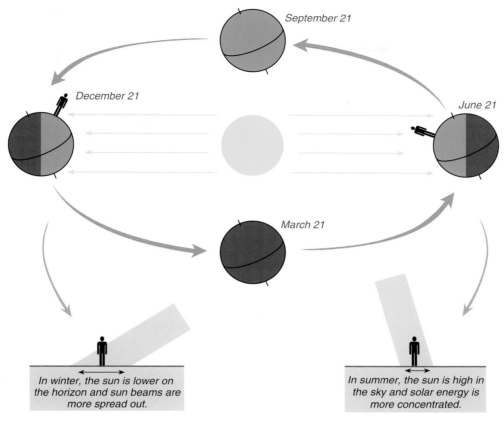

Figure 4.11. Changes in insolation as Earth orbits the sun. Example given for the northern midlatitudes.

our bucket analogy, the energy provided by the rising and setting sun is similar to the water pouring into the bucket from a faucet that is turned on during the day (increasing after sunrise, and decreasing after noon), and turned off at night. The energy loss by emission of infrared radiation is similar to the water flowing out of the hole: it is occurring all the time, night and day, and is relatively constant (within the temperature range of the surface – recall that the emission of radiation is a function of temperature). Therefore, even when the sun rises and starts warming the surface (small input), the surface is still cooling by emission of infrared radiation (large output) and might in fact be losing more energy than it is gaining (output > input). Thus, the temperature may still drop, even though the sun is already above the horizon. Later in the morning, the incoming solar radiation will be greater than the outgoing infrared radiation (input > output), and the temperature starts rising. Because this increase in temperature starts later than sunrise, we can speak of

meteorological sunrise. For similar reasons, *meteorological sunset* starts earlier than the actual sunset, when the incoming solar radiation becomes less than the outgoing infrared radiation.

To continue with the leaky bucket analogy, we note that the maximum water depth in the bucket is not reached when the input is greatest. The water level continues to rise, even when the input starts to decrease, as long as the input is still greater than the output. The water level is maximum when the input decreases sufficiently to match the output. Similarly, maximum temperatures do not normally occur at solar noon, when the input of solar energy is maximum, but rather close to meteorological sunset, when the incoming solar radiation is about to become less than the outgoing infrared radiation.

Figure 4.13 shows an example of the diurnal cycle at Amarillo, Texas, in February 2014. The temperature can be observed to swing up and down every day by about 20 °C. While the meteorological sunrise

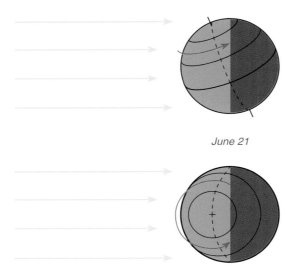

Figure 4.12. Length of day in summer: a location in the midlatitudes spends more than 12 hours on the day side of Earth (red arrow). Note the difference between the vertical line separating night from day, and the dashed line going from pole to pole and separating Earth in two hemispheres of 12 hours each. Note that the length of day is 12 hours at the equator, and 24 hours north of the Arctic Circle.

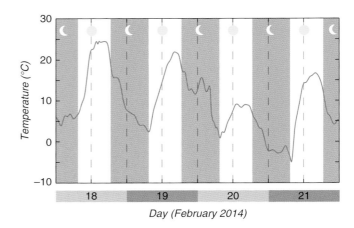

Figure 4.13. Meteogram illustrating the diurnal cycle at Amarillo, Texas, in February 2014. The red curve indicates temperature. The blue shading indicates nighttime, while the white shading indicates daytime (in local time). The blue dashed lines indicate midnight, while the red dashed lines indicate noon, for reference.

(when the temperature starts rising) here coincides closely with the actual sunrise (when the blue shading turns to white), the meteorological sunset (when the temperature starts falling) can be observed to happen *before* the actual sunset (when the white shading turns to blue).

The combination of conduction, convection, and the diurnal cycle also explains the changes observed in the temperature profile close to the surface. Figure 4.14 shows a typical temperature profile (in black) in which the temperature decreases with height. Recall that heat is transferred primarily by conduction from the surface to the first few millimeters of air, and then by convection further up, so that the atmosphere is warmer closer to the source of heat (the surface). On a sunny day, solar radiation heats the surface, which heats the first meters of atmosphere by conduction and convection (warm air rises), as shown by the red temperature profile in Figure 4.14(a). At night, the surface cools by emission of infrared radiation, which cools the lower atmosphere by conduction and yields the blue temperature profile in Figure 4.14(b).

(Note that convection does not intervene in this case, since cold air tends to sink. The air tends to settle in layers, with the coldest layers at the bottom, but there is little mixing, unless there is wind.)

Note that, in Figure 4.14, the resulting diurnal cycle is very pronounced in the first hundred meters of air, but almost non-existent higher up in the atmosphere. Indeed, conduction does not penetrate very high in the atmosphere (as we discussed earlier, air is a poor conductor and a very good insulator). Convection is more efficient, but is constrained by other processes, as we will learn in Chapter 6. As a result, changes in the temperature of the surface are felt over roughly the lowest kilometer of the atmosphere. In the absence of weather systems, the vertical temperature profile changes slowly, except for a "bottom leg" that sweeps between high and low temperatures throughout the diurnal cycle.

Note that, in the nighttime temperature profile (in blue), temperature *increases* with height in the first hundred meters of atmosphere, which is the *inverse* of the average temperature profile (in black). Recall that we call this feature a **temperature inversion**. Indeed, a nighttime temperature inversion often forms when the surface loses a sufficient amount of energy by emission of infrared energy overnight (radiational cooling).

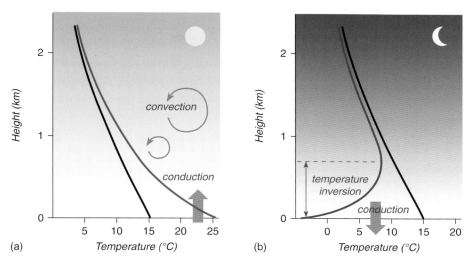

Figure 4.14. Diurnal cycle in temperature close to the surface. (a) Daytime profile. (b) Nighttime profile.

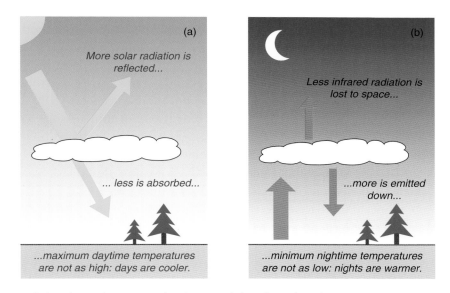

Figure 4.15. Impact of clouds on the energy budget and the diurnal cycle.

This inversion is most pronounced close to the surface. We will learn later that such a temperature inversion also makes the lower atmosphere very **stable**, which traps pollution (see Chapter 14).

4.5.4 The Influence of Clouds

The presence of clouds affects both day and night portions of the diurnal cycle: clouds prevent incoming solar radiation from heating the surface, limiting high temperatures during the day (Figure 4.15(a)), and they prevent infrared radiation from escaping to space, limiting low temperatures at night (Figure 4.15(b)). As a result, cloudy days are typically cooler than sunny days and cloudy nights are typically warmer than clear nights. This leads to a smaller diurnal cycle (i.e., narrower day–night temperature range). The climate of any given location will reflect this impact of cloudiness. Seattle, Washington, for example, has a maritime climate characterized by relatively high amounts of cloudiness. The average low and high temperatures in January are 3 and 8 °C (37 and 47 °F), respectively: a 5 °C (10 °F) diurnal cycle. By contrast, Denver, Colorado, has a more continental climate with much less cloudiness. The average low and high temperatures in January are –9 and 6 °C (15 and 43 °F, respectively: a 15 °C (28 °F) diurnal cycle.

4.5.5 Land–Ocean Contrasts

Extending the Seattle–Denver comparison, we observe that locations influenced by the ocean tend to have small diurnal cycles, while continental locations can experience much larger daily extremes. Why do land masses cool and heat so easily, while bodies of water (and their surrounding coastal lands) maintain a comparatively narrower temperature range throughout the day and throughout the year?

Water is a very special molecule with very interesting properties. The presence of one oxygen atom and two hydrogen atoms at an angle of 104° and the asymmetric distribution of positive and negative charges between the hydrogen and oxygen atoms give unique properties to the water molecule. One property you might know is the propensity for water molecules to arrange themselves into a special hexagonal pattern when they freeze, which explains the formation of delicate ice crystals and snowflakes of definite shape, and the lower density of ice compared to liquid water. Of more interest to us is the high **heat capacity** of liquid water. Heat capacity is the energy required to increase vibrational energy as the temperature increases. In the case of liquid water, however, additional energy is required to break "hydrogen bonds" – weak bonds that exist between the hydrogen atoms (positively charged) of a water molecule and the oxygen atoms (negatively charged) of *other* water molecules when they are in contact. Although weak, these hydrogen bonds are constantly broken and recreated in liquid water. They are strong enough that it requires additional energy for water molecules to speed up, as these bonds need to be loosened continually. This additional energy corresponds to the higher specific heat of liquid water. These hydrogen bonds are not broken during the warming of ice, and they are already broken in water vapor. Therefore, the heat capacity of ice and water vapor is less than that of liquid water.

To provide a more quantitative comparison, let us consider one gram of liquid water exposed to one calorie of solar energy (i.e., 4.18 joules). The gram of water will heat up by one degree Celsius. This is by design, as the calorie is defined as the amount of energy required to heat up one gram of water by one degree Celsius. We also say that the heat capacity of water is equal to:

$$c_p \text{ (water)} = 1 \text{ cal}/(g °C) = 4.18 \text{ J}/(g °C)$$

In contrast, the heat capacity of soil is about $0.2 \text{ cal}/(g °C)$:

$$c_p \text{ (soil)} = 0.2 \text{ cal}/(g °C) = 0.84 \text{ J}/(g °C)$$

It takes only 0.2 calorie to heat up the gram of soil by one degree. Therefore, if we place the soil next to the water and expose them both to 1 calorie of solar energy, the soil will heat up by 5 degrees – five times more than the water! That is why, when you go to the beach, dry sand can be burning hot while wet sand remains relatively cool.

This is one of four reasons why water temperature changes more slowly than land temperature. For completeness, we list them all here:

❶ As we just described, water has a higher heat capacity than soil.

❷ The energy absorbed by land is concentrated in the first few millimeters of soil, since conduction is not very efficient at transferring heat downward. In contrast, the surface of the ocean is stirred by the wind and mixing redistributes surface water through a much deeper layer. As water is mixed downward, heat is also transported downward and spread throughout the layer – it is somewhat similar to convection. A given amount of heat distributed throughout a thicker layer of water means lower temperatures than the same amount of heat concentrated in the first millimeters of soil.

❸ Because water is transparent to visible light, solar radiation reaches below the surface and penetrates through a deeper layer of water than it does for land. The absorption of energy is distributed through a thicker layer, and, as a result, heat is more spread out, which also results in smaller temperature changes.

❹ Part of the solar energy hitting the water surface is used to evaporate water, rather than to heat

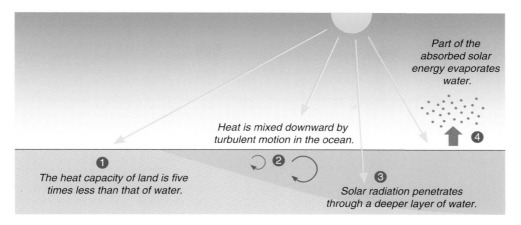

Figure 4.16. Four factors contribute to land masses heating up much more than oceans under the sun.

it (see Chapter 5). In technical terms, the energy is used as latent heat of evaporation, rather than sensible heat. We can think of this latent heat as lost and contained in the water vapor, off to a new destination in the atmosphere. Since there is less energy available to heat the water, temperatures do not increase as much as over land.

These effects contribute to creating large temperature contrasts between land and ocean. In the same way that latitudinal temperature contrasts between the tropics and the poles drive large-scale weather patterns such as midlatitude cyclones, land–ocean temperature contrasts drive smaller-scale circulations such as sea breezes and larger-scale circulations such as monsoons, which we will discuss in Chapters 8 and 9, respectively.

Summary

In the atmosphere, heat can be transferred in three different ways: conduction, convection, and radiation.

- ▶ **Conduction** is a process by which heat is transferred by contact, collision, and transfer of molecular vibrations, which is slow and inefficient in the atmosphere – air is a poor conductor.

- ▶ **Convection** is a faster and more efficient process by which air is displaced vertically, along with its heat content, due to

temperature and density differences – "warm air rises" due to buoyancy. Pockets of warm air rising by convection are called **thermals**.

- ▶ **Radiation** carries electromagnetic energy that can be reflected, transmitted, scattered, or absorbed by the different elements of the Earth system.

Radiation takes the form of **electromagnetic waves** characterized by a **wavelength** and a frequency. Lower frequency electromagnetic waves, such as **infrared** radiation, have longer wavelengths and carry less energy than visible light. Higher frequency electromagnetic waves, such as **ultraviolet** radiation, have shorter wavelengths and carry more energy than visible light.

When electromagnetic radiation is **absorbed** by a material, it is converted to heat, and the temperature of the material increases.

The reflective properties of a material are defined by its **albedo**, which is the percentage of incoming sunlight that is **reflected** by the object or surface. The albedo of snow can be as high as 90%, whereas the albedo of the oceans and forests is as low as 10%.

Earth also **emits** electromagnetic radiation, mostly in the infrared. Earth is close to **radiative equilibrium**, by which the amount of incoming solar energy that is absorbed is equal to the amount of outgoing infrared energy emitted by Earth.

Certain variable gases of the atmosphere, such as water vapor, carbon dioxide, and ozone, absorb and emit radiation at specific wavelengths.

By absorbing and reemitting infrared radiation, some of these gases contribute to the **greenhouse effect** and maintain Earth's surface at an average surface temperature of 15 °C, 33 °C warmer than without the atmosphere. The enhancement of the greenhouse effect by anthropogenic increase of certain greenhouse gas concentrations, such as carbon dioxide and methane, leads to **global warming** and climate change.

The angle of sun rays at different latitudes and the albedo of snow and ice at the poles create a net warming at the equator and a net cooling at the poles. The resulting **temperature contrast** generates currents in the ocean, and air motion in the atmosphere to move heat poleward, including cyclones in the midlatitudes.

The tilt of Earth's axis exposes each hemisphere to more solar radiation in summer and less solar radiation in winter, resulting in **seasonal** variations in temperature. The tilt of Earth's axis also exposes each hemisphere to longer days in summer and shorter days in winter, which enhances the seasonal contrasts.

The daily rotation of Earth causes **diurnal variations** in temperature. The diurnal cycle is most pronounced near the surface, due to the absorption and emission of radiation by the ground, conduction of heat between the ground and the atmosphere, and convection. At night, the decrease in temperature at the surface caused by emission of infrared radiation can create a **temperature inversion**, where temperature increases with altitude.

The presence of clouds in the atmosphere limits the amount of solar radiation that reaches Earth's surface during the day, and limits the emission of infrared radiation out to space during the night, which reduces the amplitude of the diurnal cycle.

Water bodies experience smaller temperature variations than land surfaces due to turbulent mixing, deeper penetration of sunlight, the evaporation of water, and a higher heat capacity. Therefore, the overlying air and coastal locations also experience a smaller diurnal and seasonal cycle than inland, continental locations.

CHAPTER 5

Water

Water in its different forms creates the possibility of a wide array of atmospheric phenomena, from clouds to rainbows, and from lightning to hurricanes. Before we can describe these phenomena in detail, however, we need to understand how and when water can change phase, from solid (ice), to liquid (liquid water), and gas (water vapor). To that end, we will explore different aspects of the water cycle, in particular condensation and evaporation, leading to the formation of clouds and precipitation. We will define quantities of great importance to this exploration: saturation, relative humidity, and dew point temperature.

CONTENTS

The presence of water on Earth makes our planet very special. Most importantly for us, it makes life possible. But the fact that water can exist in three different phases, or states, in the same environment (solid, liquid, and gas) makes the Earth system even more special: it provides additional paths for efficient energy transfer. Thus, to understand the role played by water in the atmosphere, we first need to understand how water cycles through the Earth system in these different states.

5.1 The Water Cycle

On Earth, water can exist under three forms, or **phases**: solid (ice, hail, ice crystals, frost, snowflakes), liquid (ocean water, cloud droplets, rain), and gas (water vapor). These phases differ in the arrangement of water molecules, as well as in kinetic energy. In solid ice, for example, molecules do not move around at all, but merely vibrate, which corresponds to a lower energy state. Moving molecules in

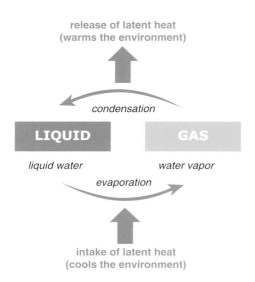

Figure 5.1. Intake and release of latent heat as water changes phase between liquid and gas.

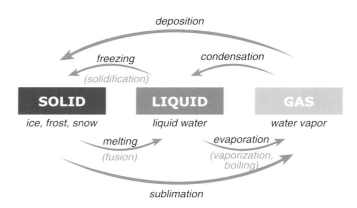

Figure 5.2. Phase changes of the hydrologic cycle with corresponding intake (blue arrows) and release (red arrows) of latent heat. (Alternative terms used in physics are indicated in gray.)

liquid water and water vapor contain comparatively more energy. More importantly, for ice to melt and for water to evaporate, the hydrogen bonds existing between water molecules (see Chapter 4) must be broken, which requires additional energy. Therefore, to change state from solid to liquid or liquid to gas, water molecules need an energy boost: enough energy to break the hydrogen bonds. This extra boost of energy could be obtained from the sun, for example. Conversely, when water returns from a higher energy level to a lower energy level, from water vapor to liquid (condensation), or from liquid to ice (freezing), this extra boost of energy is released (Figure 5.1). In other words, it is returned to the surrounding environment (to other gases, for example, such as nitrogen and oxygen). In that respect, we can think of water vapor as containing an extra quantity of energy that allows it to exist in the vapor phase: we call it **latent heat**, as if it is in a latent state, waiting to be released. We will see that the release of latent heat is of critical importance, as it can effectively warm the environment. Conversely, when water evaporates, or when it changes phase directly from solid to vapor (**sublimation**), it takes the necessary latent heat from the environment, which tends to *cool* the environment. (It cools the ocean, for example, in the case of evaporation of water from the ocean.)

When water returns from the liquid to the solid state (freezing), latent heat is also released. When it goes straight from the vapor phase to the solid phase (**deposition**), as in the formation of frost or snow flakes, an even greater amount of latent heat is released, since this process is the equivalent of two phase changes. This gives us the schematics shown in Figure 5.2, in which all phase changes are summarized, along with the gains or losses of latent heat.

Let us now investigate the conditions under which these phase changes are possible.

5.2 Saturation

If you have ever forgotten a glass of water on a table for several days, you know that water tends to evaporate over time. After a few days, the glass will be empty and dry. If you cover the glass with a lid, however, the water will not completely evaporate. Even if there is only a little bit of water, and a lot of space in the glass, the water will never completely evaporate to occupy that space as water vapor. It is quite an important observation, as it suggests that there is a limit to the amount of water that can be evaporated in a given space, or volume. Indeed, if we imagine an open water surface under dry air, the water will first evaporate into the dry volume, as individual molecules escape the liquid phase to become water vapor (Figure 5.3(a)). As the amount of water vapor molecules increases, however, some water molecules will start rejoining the liquid surface underneath, as

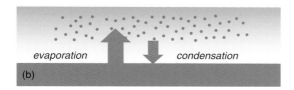

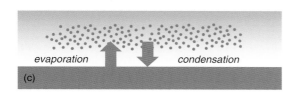

Figure 5.3. From evaporation to saturation. (a) Water evaporates into dry air. (b) Evaporation is greater than condensation, so there is positive net evaporation. (c) When saturation is reached, condensation balances evaporation.

a result of random motion and collisions. Thus, condensation is taking place at the same time as evaporation continues (Figure 5.3(b)). As long as evaporation is greater than condensation, or, in other words, as long as the *net* evaporation is positive, the amount of liquid water will diminish. At some point there will be enough water vapor above the liquid water surface that the condensation flux will equal the evaporation flux: a state of equilibrium in which evaporation and condensation are continuously taking place while the overall amount of liquid water and water vapor does not change (Figure 5.3(c)). This is yet another example of dynamic equilibrium (see Appendix 3.1) and is somewhat similar to our bucket of water with equal input and output, or to a room in a museum in which the rate of people entering the room equals the rate at which other people are leaving the room, so that the total number of patrons enjoying the art in the room remains constant, even though the patrons themselves are never the same.

When we have reached that state of equilibrium, and therefore the maximum amount of water vapor

that can occupy the space above the liquid water, we say that the air is **saturated** with water vapor, or that we have reached **saturation**. [In reality, the presence of dry air is irrelevant, as gases behave independently of each other, and it is really the space, or volume, that is saturated with water vapor. We would obtain the same result in a vacuum. But since we always deal with air in meteorology, it is common practice to say that the *air* is saturated.]

Note that the notion of saturation is relative to the surface available for the water vapor to condense upon. In the above example, we considered a flat surface of water. But water vapor typically condenses upon many different types of surface. For example, grass provides a useful surface for **dew** to form, while in the atmosphere water vapor will condense upon available solid particles called **aerosols**, such as dust particles and ice crystals. We will see shortly that the amount of water vapor in the air at saturation will depend upon the available surface types.

We can foresee that these concepts will have important implications, as condensation and cloud formation must depend on that level of saturation. In fact, lacking a suitable surface to condense upon, water vapor can reach much higher concentrations than we would find otherwise. Before we go any further, however, let us introduce another useful notion.

5.3 Humidity

If we are to discuss the amount of water vapor present in the air, or humidity, to understand the formation of clouds and rain, we need a way to quantify it, and therefore a way to measure it. Instruments that measure humidity are called **hygrometers**. We can resort to different methods to estimate the amount of water vapor around us, each leading to a different measure and different units. We can express it as the mass of water vapor per volume of air (in which case it is called **absolute humidity**), as the mass of water vapor per mass of air (**specific humidity**), or the mass of water vapor per mass of "dry" air, i.e., per mass of all other gases except water vapor (**mixing ratio**). In our case it will be more interesting to use yet another measure called **water vapor pressure**: the pressure exerted in a given space by water vapor

Box 5.1. Dalton's Law

In 1801, John Dalton (1766–1844) established that, in a mixture of gases, the total pressure exerted by the gas mixture is equal to the sum of the partial pressures exerted by the individual gases. We usually refer to this statement as "Dalton's law".

For example, imagine an air parcel containing nitrogen, oxygen, and water vapor with respective partial pressures p_{N_2}, p_{O_2}, and p_{H_2O}. Dalton's law states that the total air pressure of the parcel is equal to

$$p = p_{N_2} + p_{O_2} + p_{H_2O}$$

For many applications, it is useful to differentiate the partial pressure of dry air (p_{dry}) and the partial pressure of water vapor (usually written 'e'):

$$p = p_{dry} + e$$

Since air pressure is a variable that we are already using, expressing the water vapor content of the air as water vapor pressure is convenient, and simplifies some equations related to weather physics.

alone, as if all other gases were removed from that space (see Box 5.1). We use the lowercase letter *e*, and, as for atmospheric pressure, we will use hectopascals (hPa) as our unit. In this unit system, water vapor pressure (*e*) is typically on the order of ones to tens of hectopascals.

These measures of humidity, however, are not necessarily always helpful to us. How is a water vapor pressure of 3 or 30 hPa relevant to the formation of clouds and rain? A more interesting concept arose as we were discussing evaporation and condensation: that of saturation.

5.4 Relative Humidity

Have you ever noticed that you are really cold when you come out of the swimming pool, even when the air is warmer than the water? Indeed, when you come out of the pool, you are usually covered with water, and this water evaporates into the air, which requires energy. The water needs an extra boost of energy to leave the liquid phase, so that the molecules of water can detach and turn into water vapor (Figure 5.2). It needs *latent heat of evaporation*. This latent heat is taken from your body, which cools you down.

Similarly, when playing sports, your body cools by *sweating*. Indeed, remember that humans have evolved to sweat because evaporation of sweat into the surrounding air is an efficient cooling mechanism. In fact, it cools you down 100 to 1000 times faster than if you were to wait for conduction to transfer your

excess heat away by contact, through molecular transfer. [Some researchers think that the evaporative cooling of sweat is one of the characteristics acquired by humans two million years ago that allowed us to hunt other animals by pushing them to heat exhaustion, i.e., endurance running and persistence hunting.]

Now, if you are playing in a dry environment, your sweat will evaporate fairly easily, as the air can still contain additional water vapor. If you are playing in a moist environment, such as on a humid day, for example, the air already contains a lot of water vapor, and there is less evaporation potential. If the air is *saturated* with water vapor, it is especially difficult to evaporate sweat into it. The maximum has been reached, so to speak. Sweat remains on your skin, your body keeps heating up, and you might suffer heat stroke.

We could return to the swimming pool and observe similar extremes. If you come out of the water to find yourself in a steamy Turkish bath, saturated with water vapor (as opposed to a hot, but dry, sauna), you will not evaporate much water, and therefore will not cool effectively. If you come out of an outdoor pool in a hot, dry place, such as in a desert area, evaporation will be so strong that you will feel quite cold before the sun starts warming you again.

Thus, more than the actual amount of water vapor in the air, it is the amount of water vapor in the air *compared to the maximum there can be* that is of interest to us. In other words, how close we are to saturation. If we are far from saturation, evaporation is stronger. If we are close to saturation, evaporation is limited – and more importantly for us, as we will see shortly, condensation is not far from happening, which creates the possibility of clouds.

This leads us to define a new quantity that will be much more useful to us. We first define the water vapor pressure that the air would have if it were saturated: the **saturation water vapor pressure** (e_s), where the subscript stands for "saturation." And we further define the percentage of water vapor present in the air, compared to the maximum amount at saturation, as the **relative humidity** (RH):

$$RH = \frac{\text{actual water vapor pressure}}{\text{saturation water vapor pressure}} = \frac{e}{e_s}$$

For example, if the maximum amount of water vapor that can be present in the air is $e_s = 12\,\text{hPa}$ (we can think of it as the maximum capacity), and the actual amount that is present is only $e = 9\,\text{hPa}$, then the relative humidity is

$$RH = 9/12 = 0.75 = 75\%$$

Note, also, that the above formula can be manipulated to become

$$e = RH \times e_s$$

That is, if we know the relative humidity and the saturation vapor pressure, we can calculate the amount of water vapor as measured by the vapor pressure, which can be handy in some circumstances (see section 5.5.1).

Since this measure of humidity is only relative to a maximum capacity, but not absolute as in the mass of water per volume of air, we call it *relative* humidity. This subtlety is important to remember, as the relative humidity can be high even though there is very little moisture in the air, as we will illustrate shortly with an example.

Note that, in rare circumstances, the air can contain *more* than its maximum capacity: we say that the air is *supersaturated*. This is the case, in particular, in the absence of a surface for the water vapor to condense upon. Recall that we have defined saturation with respect to a flat surface of water. The surface provides a stable *anchor point* for the water vapor molecules to attach to. When there is no surface, however, water vapor molecules bump into each other, and have nothing to attach to. Arguably, they could attach to each other, but since they have high kinetic energy, they tend to push each other apart after each collision. For them to form an initial *clump* of water molecules, a much higher RH than 100% is required, so that the water vapor pressure would be sufficient to keep the water molecules together. While this process, called **homogeneous nucleation**, is possible in theory, it rarely happens in practice, because the atmosphere is full of small, invisible, solid or liquid particles called **aerosols**, on which water vapor can readily attach, as we will see shortly.

5.5 Humidity and Temperature

We can easily think of circumstances in which the water vapor pressure will change: if we evaporate water from the ocean, for example, the amount of water vapor in the air (e) will increase, and the relative humidity (RH) will increase as a result. But can the saturation water vapor pressure change? Can the maximum capacity of the air change under any circumstances? Although this is more difficult to conceive, a little thought experiment will help.

Let us return to our open water surface under saturated air (Figure 5.3(c)). We had reached a state of equilibrium in which the upward evaporation flux was balanced by a downward condensation flux. Let us imagine that the whole system warms for some reason: there is overall more energy in the water, and more energy in the air. The whole system is at a higher temperature. Having more energy, the liquid water molecules will tend to evaporate more easily, which will increase the evaporation flux. Having more energy overall, these water vapor molecules will tend to remain as vapor rather than return to the liquid phase, which increases the total amount of water vapor in the air. Shortly thereafter, the system will reach a new state of equilibrium in which condensation once again balances evaporation, but the system can now sustain more water vapor in the gas phase: we have increased the *capacity* of the air. (To return to our museum analogy, it is now a busy weekend, we

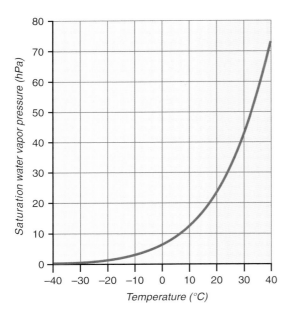

Figure 5.4. Saturation water vapor pressure with respect to liquid water in relation to temperature (Clausius–Clapeyron formula).

have more visitors coming into the room, more visitors walking out, and therefore more visitors in the room at any time.)

Thus, in short, warmer air can contain more water vapor. [Note again that, strictly speaking, the air itself has nothing to do with it. It is merely the water that can more easily exist as vapor due to the higher temperature, independent of the presence of warm air. But it is common parlance to say that "warmer air can contain more water vapor."] Conversely, the colder the air, the less water we can evaporate into it. This is captured mathematically by a formula derived in the nineteenth century by Émile Clapeyron (1799–1864) and Rudolf Clausius (1822–1888), and represented graphically in Figure 5.4.

Note that the increase in saturation water vapor pressure with temperature is exponential, so that the rate of increase is much greater at high temperature than at low temperature. For example, a 5 °C temperature increase from 35 to 40 °C corresponds to an increase in saturation water vapor pressure of 17 hPa, whereas a 5 °C temperature increase from 0 to 5 °C corresponds to an increase in saturation water vapor pressure of only 2 hPa. This has very important consequences on Earth, where we have warm, humid tropics and cold, dry poles.

5.5.1 Relative vs. Absolute Humidity

Two applications come to mind. First of all, temperature changes with latitude and environmental conditions. Since the saturation water vapor pressure (e_s) is a function of temperature, we can also expect it to vary with latitude and environmental conditions, along with relative humidity (RH), which contains e_s in the denominator. For example, if the temperature is –20 °C in Canada, we can read in Figure 5.4 that the saturation water vapor pressure e_s is equal to 2 hPa. If the actual water vapor content is 2 hPa, then the relative humidity will be

$$RH = e/e_s = 2/2 = 1 = 100\%$$

In contrast, if the temperature is 40 °C in Death Valley, California, we read in Figure 5.4 that $e_s = 74$ hPa. Even if the relative humidity is only 40%, which is much less than the case before and very far from saturation, the actual water vapor content is

$$e = RH \times e_s = 0.4 \times 74 = 30 \text{ hPa}$$

In other words, even though the relative humidity is low, the actual water vapor content (30 hPa) is much higher than in the previous case (2 hPa), where the relative humidity is 100%. This reinforces the distinction between absolute and relative humidity. Even though both are measures of humidity, one can be high while the other is low. They tell us two very different things about moisture. Absolute humidity tells us about the actual amount of water vapor in the air, whereas relative humidity tells us how close or far the air is from saturation.

5.5.2 Condensation

The second application that comes to mind is of critical importance for understanding the formation of clouds. Let us imagine a parcel of air at 20 °C containing 15 hPa of water vapor. It is close to the ground and, as the sun sets and the emission of infrared radiation by the ground becomes greater than the absorption of solar radiation, the ground starts cooling. By conduction of heat from the overlying air to the ground, the air parcel starts cooling as well.

Table 5.1. *Example of saturation and condensation by temperature decrease.*

T (°C)	e_s (hPa)	e (hPa)	RH (%)
20	23	15	65
19	22	15	68
18	21	15	72
17	19	15	75
16	18	15	83
15	17	15	88
14	16	15	94
13	15	15	100
12	14	14	100
11	13	13	100
10	12	12	100

Figure 5.5. Dew forms on blades of grass as the temperature drops and the air becomes saturated.

At first, with a temperature of 20 °C, we can read in Figure 5.4 that the saturation water vapor pressure (e_s) is equal to 23 hPa, and since the actual water vapor pressure (e) is 15 hPa, we can calculate that the relative humidity is

$$RH = e/e_s = 15/23 = 0.65 = 65\%$$

(This and all subsequent results are summarized in Table 5.1.)

As the air parcel cools, however, its water vapor *capacity* decreases. At 19 °C, for example, the saturation water vapor pressure is only 22 hPa, and the relative humidity is now

$$RH = e/e_s = 15/22 = 0.68 = 68\%$$

In other words, as temperature decreases, the relative humidity *increases*. [Note that e_s is in the denominator, and therefore we have a relationship of *inverse proportionality*.] Table 5.1 shows how the relative humidity slowly increases to 100% as the temperature decreases to 13 °C as the night progresses. Pause for a second to ponder this result, as it is critical for our understanding of cloud formation later on. Note that the water vapor content (e) of the air parcel is

not changing. It remains constant at 15 hPa. Thus, we are not saturating the air parcel by adding water vapor into it: we are getting closer to saturation by *decreasing the water vapor capacity of the air parcel*. To return to our museum analogy, we could imagine a room occupied by a few visitors whose number does not change. The room is not full, but the size of the room decreases, as if the walls were closing in on the visitors. At some point, when the capacity of the room decreases to the exact number of visitors, the room will become full (saturated), but it is very different from adding visitors to the room. Here in our air parcel, we have decreased the capacity to 15 hPa, which is the exact amount of water vapor we had in the parcel to start with.

What happens if we decrease the temperature further? The water vapor capacity will decrease further and will be less than the actual amount of water vapor in the air parcel. At 12 °C, the capacity decreases to 14 hPa, but the air parcel contains 15 hPa. We have too much water vapor, and 1 hPa must leave the gas phase to become either liquid or solid. In other words, 1 hPa of water vapor must condense (or be deposited as ice if the temperature is below freezing). Typically, close to the ground, water will condense on blades of grass, flowers, spider webs etc., to form **dew** (Figure 5.5). At subfreezing temperatures, it will form **frost** (Figure 5.6). As the temperature keeps

Figure 5.6. Frost forms when the air becomes saturated below 0 °C.

dropping throughout the rest of the night, the saturation water vapor pressure keeps decreasing, and more water vapor has to leave the gas phase to condense and form dew. This process will be the motivation for introducing yet another measure of humidity, the dew point temperature – see Section 5.6.

Note that, beyond this point, the actual water vapor pressure (e) decreases along with e_s, because more and more water vapor leaves the gas phase. Therefore, the relative humidity remains at 100%.

Two other interesting phenomena can be observed at this point. First, as the ground gets colder and conduction proceeds, a deeper layer of air might cool down to saturation and dew might be observed higher up above the ground, on tree branches and rooftops. Second, water vapor might start condensing on aerosols present in the air (suspended solid particles such as salt, dust, pollen, smoke, pollution, etc.) to form tiny water droplets. In other words, we might observe **fog**. This kind of morning fog is very common after clear nights when the ground has had a chance to cool by emission of infrared radiation. (Recall the temperature inversion of the nighttime temperature profile in Figure 4.14). It is typically lying low and close to the ground (where conduction is most efficient) and is appropriately called **radiation fog**.

We will return to the formation of cloud droplets in chapter 6.

5.6 Dew Point Temperature

We now have everything we need to introduce our last measure of humidity. In Section 5.5.2, we demonstrated how condensation could be predicted to happen if, without changing the pressure or the water vapor content of the air, we decreased the temperature of the air (by removing heat) down to a point where the saturation water vapor capacity (e_s) becomes equal to the actual water vapor pressure (e). Specifically in our example, we found that by decreasing the temperature to 13 °C, we reach saturation, and by cooling the air *below* 13 °C, condensation starts. Because dew starts forming at 13 °C, we call that threshold temperature the **dew point temperature** (T_d), or "dew point," for short.

Note that the dew point temperature is not a temperature that *must* be reached, although it *can* be reached. It is a theoretical threshold that informs us about the water vapor present in the air, and that is reached only when and if the air is cooled down sufficiently at constant pressure. If there is no change in temperature (or moisture), the dew point temperature is never reached.

Note further that the dew point temperature is specific to the air parcel and depends on the specific conditions of the parcel. You can verify for yourself that, if the air parcel contained 18 hPa of water vapor to start with, the dew point temperature would be 16 °C. If the air parcel contained only 12 hPa of water vapor, the dew point temperature would be 10 °C. The dew point temperature, in spite of its name, does not directly tell us anything about the temperature, or heat content of the air parcel (even though it is called a "temperature"). It provides useful information about the *moisture content* of the air parcel.

Thinking backward, we can take advantage of the previous observation. If the air parcel, still at 20 °C, is very moist (for example, e = 22 hPa), the dew point temperature is high and close to the actual temperature (T_d = 19 °C, which is close to T = 20 °C). In other words, we need only a small drop in temperature to reach saturation. If the air parcel is dry (for example, $e = 12\,\text{hPa}$), the dew point temperature is low and far from the actual temperature ($T_{\text{d}} = 10\,^{\circ}\text{C}$, which is further from $T = 20\,^{\circ}\text{C}$). In other words, we need a large temperature drop to reach saturation. This temperature drop to the dew point is a useful concept in itself, called the **dew point depression**, i.e., the drop in temperature necessary to reach saturation, $T - T_{\text{d}}$.

Therefore:

> If T_{d} is high and close to T, the air is moist.
>
> If T_{d} is low and far from T, the air is dry.

Or, alternatively:

> small dew point depression = moist
>
> large dew point depression = dry.

Thus, the dew point temperature is really a measure of humidity. Like the relative humidity, it tells us how far we are from saturation, but it circumvents the need to measure the actual water vapor pressure (e). This has technological advantages, because measuring e is challenging. Measuring T_{d}, however, can be done with a trick. By cooling a mirror, we can observe when the mirror fogs up with condensation, which indicates that we have reached the dew point. This can be automated in a weather station by shining a beam of light on the mirror while it is being cooled. As long as the mirror is clear, the beam will be reflected, which can be detected with a sensor. As soon as the beam stops being reflected (because of the condensation), we have reached saturation and T_{d} can be recorded.

Finally, note that the dew point temperature also informs us about what will happen if the temperature continues to decrease after saturation has been reached. Recall that latent heat is released when water changes from the gas to the liquid phase, as dew forms. This release of latent heat warms the air, often reducing the rate of cooling after reaching saturation. At warm temperatures, when there is a lot of water vapor in the air at saturation, this provides a very powerful limit on how cold it can get overnight: the temperature will drop to near the dew point, and then stop decreasing once dew forms, due to the release of latent heat. So the dew point temperature can be a useful guide to morning low temperatures under warm, clear, and calm conditions. At very cold temperatures, there is so little water vapor in the air at saturation that there is little latent heat release; as a result, the dew point is not a good guide to morning low temperatures when the air is very cold.

5.7 Applications of the Dew point Temperature

The dew point is a common measure in meteorology, and is used in various situations. Three examples will further reinforce this new concept.

5.7.1 Surface Weather Maps

Figure 5.7(a) shows the surface weather map corresponding to our February 2014 cyclone, now indicating dew point temperature measurements. It is common practice to indicate the actual temperature in the upper left corner and the dew point temperature in the lower left corner of the data for each weather station. By looking at individual weather stations, we can appreciate how it is warmer and moister closer to the Gulf of Mexico (southeast), and colder and drier further west and north. To appreciate the distribution better, we have colored and contoured the dew point temperature field in Figure 5.7(b), and reproduced our temperature and pressure surface map in Figure 5.7(c). We see the cold and warm fronts, the warm sector, and the low pressure center reflected in the dew point temperature field (see Chapter 2).

It appears in Figure 5.7(b) that the warm sector (the sector of warm air delimited by the two fronts) is also particularly moist (high dew point temperature). It makes sense if we match the measurements with the wind direction in the warm sector: this warm and moist air is essentially blown inland from the Gulf

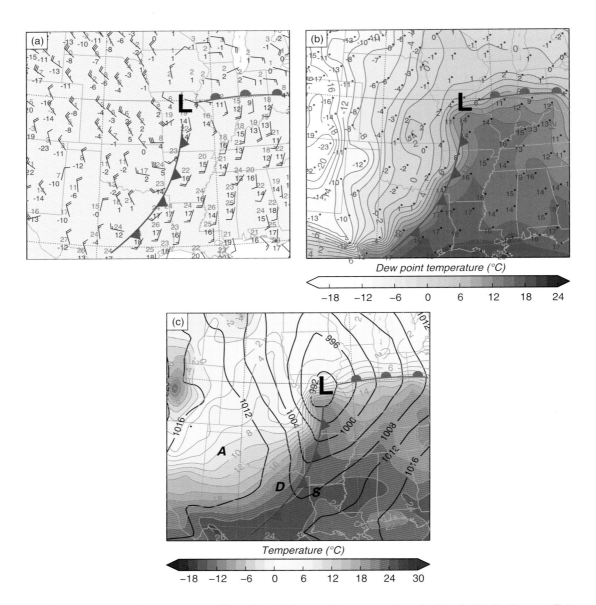

Figure 5.7. (a) An example of temperature (in red) and dew point temperature (in blue) distribution on February 20, 2014, at 18:00 UTC. (b) Corresponding dew point temperature field, colored and contoured. (c) Corresponding temperature and pressure field. Fronts and low pressure center are indicated for reference. The location of Amarillo, Texas (**A**), Dallas, Texas (**D**), and Shreveport, Louisiana (**S**) are indicated in (c) to aid help discussion. See the text for details.

of Mexico, i.e., from the warm and moist subtropics. By contrast, the wind in the cold air mass behind the cold front is northwesterly and brings dry, cold air from continental Canada. Therefore, the dew point is much lower and there exists a zone of sharp transition from low to high dew point where the contours are tightly packed together. In this particular situation, the zone of transition roughly coincides with the cold front. Thus, the frontal region, which separates cold and warm air with a strong temperature gradient, also separates dry and moist air with a strong *moisture gradient*.

Let us now explore how all this translates to a single weather station.

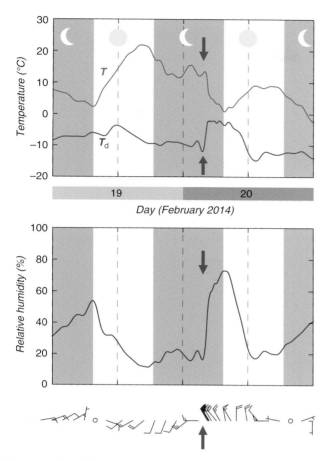

Figure 5.8. Meteogram of temperature, dew point temperature, relative humidity, and wind at Amarillo, Texas, on February 19–20, 2014 (local time). The passage of the cold front is indicated by blue arrows.

5.7.2 Meteograms

When measured at a station and plotted against time, along with temperature and wind, the dew point tells us about local changes in humidity, in particular as a response to the horizontal displacement of moist or dry air.

Figure 5.8 shows a meteogram indicating temperature, dew point temperature, relative humidity, and wind at Amarillo, Texas, on February 19–20, 2014, during the passage of our February 2014 cyclone. [The location of Amarillo is indicated by A in Figure 5.7(c) for reference.] Each wind barb at the bottom of Figure 5.8 corresponds to a wind measurement. Thus, it is a time series of wind speed and

direction at the weather station over the course of the two days displayed in the meteogram. The wind can be observed to switch direction from southwesterly to northwesterly, and to increase in speed, early on February 20, when the cold front passes through the city (see blue arrows in Figure 5.8). We will learn in Chapter 10 that extratropical cyclones typically move from west to east. Thus, if we recall Figure 5.7(a) and imagine that the whole wind field was further west at an earlier time, and then swept eastward through Texas, we can see that the weather station is first in the warm sector, where the wind is southwesterly, then in the frontal region, and finally behind the cold front, where the wind is northwesterly, consistently with the meteogram. Finally, we will learn that the wind is typically stronger around a cold front, which is also consistent with the wind speed observations in Figure 5.8.

Looking at temperature, we again find some signs of the diurnal cycle discussed in Chapter 4, but the cycle is disrupted by the passage of the cold front. Before we analyze temperature in more detail, however, let us look at humidity. First of all, note how the relative humidity is low when T and T_d are far apart, and high when T and T_d are close to each other. That is consistent with our rule of thumb:

T and T_d far apart	= far from saturation
	= lower RH
T and T_d are close	= close to saturation
	= higher RH

There is a more interesting story in this meteogram, however. The dew point temperature can be observed to be relatively constant, except early on February 20, around 4:00 a.m., when it increases by 10 °C in less than an hour. That is when the cold front of our February 2014 cyclone moves through the city (as confirmed by the changes in wind speed and direction). Recall that a high dew point means more water vapor. Indeed, moisture, clouds, and rain tend to concentrate along temperature fronts. (See also the tongue of higher humidity along the cold front in Figure 5.7(b)) Furthermore, a higher dew point means that we are closer to saturation, and the relative humidity should increase, which is indeed what

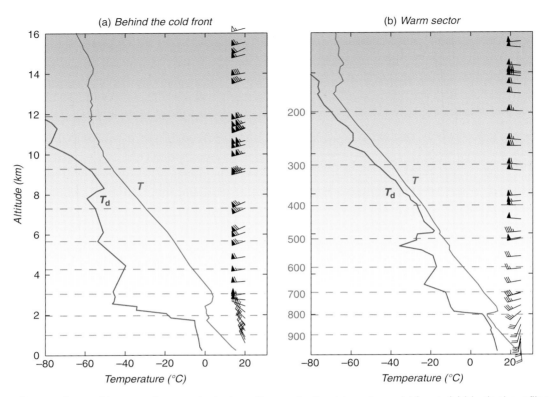

Figure 5.9. Comparison of temperature and wind profiles on both sides of a cold front. (a) Vertical profiles at Dallas' Texas, on February 21, 2014, at 00:00 UTC, on the western side of the cold front, after the front has moved through. (b) Vertical profiles at Shreveport, Louisiana, on February 20, 2014, at 12:00 UTC, on the eastern side of the cold front, in the warm sector, before the front moves through. Pressure levels (hPa) are indicated by blue dashed lines.

we observe in the lower panel of Figure 5.8 during the frontal passage.

The increase in relative humidity, however, is not only due to the increase in moisture content. A frontal passage also brings a significant change in temperature. Recall that the cold front is the boundary of a cold air mass moving into warmer air. Therefore, as the cold front was approaching from the west, Amarillo was successively in the warm air mass, then in the frontal area, and finally in the cold air mass advancing eastward, which explains the temperature drop observed between 4:00 and 7:00 a.m. The temperature drops by more than $10\,^{\circ}$C and never fully recovers during the following diurnal cycle. [Here again, looking at Figure 5.7(c) and imagining the temperature field (i.e., the cyclone) sweeping from

west to east will help you to visualize the two air masses sweeping over Amarillo and the temperature dropping when the front moves through.]

As the temperature drops, the saturation water vapor pressure also decreases, and the relative humidity increases (as in Table 5.1). Thus, the reason why the relative humidity rises so drastically between 4:00 and 7:00 a.m. (from about 20 to 70%) is twofold: the moisture content increases *and* the temperature drops. Later on during the day, northwesterly winds bring in dry air from the continental interior and the dew point temperature drops by $10\,^{\circ}$C. Furthermore, the temperature increases slightly (although not as much as the previous days), and, as a result, the relative humidity drops back down to 20%.

5.7.3 Radiosonde Profiles

The contrast between the cold and warm air masses, on each side of the cold front, can be further illustrated by studying radiosonde profiles at appropriate locations. Figure 5.9 shows two such profiles, at Dallas, Texas, in the cold air mass *after* the frontal passage, and at Shreveport, Louisiana, in the warm sector *before* the frontal passage. [The locations of Dallas and Shreveport are indicated in Figure 5.7(c) for reference.] Figure 5.9 also shows the dew point temperature, along with wind barbs at various altitudes. Each wind barb is to be read as you would on a surface map, i.e., from the north when it is pointing down, and from the south when it is pointing up. Thus, for example, the surface wind at Shreveport is not pointing upward, but *northward*, i.e., it is a southerly wind. Similarly, all upper-level winds are westerly.

By applying our now familiar rule-of-thumb, we can see that the air is much drier in the cold air mass (T far from T_d) than in the warm air mass (T close to T_d). Furthermore, the winds are most different in the lower layer of the troposphere, up to about 800 hPa. They are mostly northerly or northwesterly at Dallas, and mostly southerly or southwesterly at Shreveport. Higher up they are mostly westerly throughout the rest of the troposphere. The lower-level winds are consistent with what we observed in Figure 5.7(a), as well as with the wind shift observed in the meteogram in Figure 5.8. The drier profile at Dallas reflects the dry, cold air moving southward from the interior United States and continental Canada (northwesterly winds), while the moist profile at Shreveport reflects the warm, moist air moving northward from the Gulf of Mexico (southwesterly winds).

5.7.4 Back to Relative Humidity

Let us consolidate our new understanding with one more exercise. We mentioned that T_d was easier to measure than the water vapor pressure (e), and that the vapor pressure (e) can be recovered from T_d, which we shall now demonstrate.

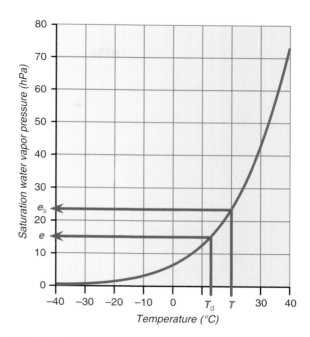

Figure 5.10. Using T and T_d to calculate e_s and e, and therefore RH $= e/e_s$.

Let us return to our air parcel and the results of our thought experiment compiled in Table 5.1. Let us further imagine that we know only the temperature of the air parcel ($T = 20\,°C$) and its dew point temperature ($T_d = 13\,°C$). We already know that we can use Figure 5.4 to deduce the saturation water vapor pressure (e_s) from T ($e_s = 23$ hPa). What does the dew point tell us? It tells us that if we cool the air parcel to 13 °C, the new water vapor capacity will be equal to the actual water vapor pressure ($e_s = e$) and the air parcel will be saturated. We can also use Figure 5.4 to calculate this new saturation water vapor pressure. At 13 °C, $e_s = 15$ hPa. (Note that this is not the original saturation water vapor pressure at 20 °C. It is a *theoretical* saturation water vapor pressure that we *would* have *if* we cooled the air parcel down to 13 °C.) Since e_s would be equal to e at this new temperature, we can conclude that the saturation water vapor pressure at 13 °C (the dew point temperature) is in fact the same as the actual water vapor pressure at 20 °C. In other words, the dew point temperature (T_d) tells us how much water vapor our air parcel contains.

These two results are summarized in Figure 5.10. A temperature T of $20\,°C$ tells us that $e_s = 23\,hPa$. A dew point temperature of $T_d = 13\,°C$ tells us that $e = 15\,hPa$. Therefore, the relative humidity is

$$RH = e/e_s = 15/23 = 0.65 = 65\%$$

and we were able to calculate RH completely indirectly, without measuring either e or e_s.

5.7.5 How to Saturate

To conclude, let us summarize and describe three different ways to saturate an air parcel.

(1) The first way is by putting more water vapor into the air parcel, as would happen by evaporation over a warm water body. It amounts to forcing e to increase to equal e_s. This is the basic principle behind **evaporation fog** (see Chapter 6).

The two other ways consist in reducing the saturation water vapor pressure e_s by decreasing the temperature. It amounts to forcing e_s to decrease and become equal to e by letting the temperature decrease to the dew point. And we can do so in two different ways:

(2) We can remove heat from the air parcel. This is the basic principle behind **radiation fog**, as described in Section 5.5.2, when the temperature of the ground decreases due to radiational cooling and heat is conducted from the air parcel to the ground. Such a process, where heat is exchanged between the air parcel and the environment, is called a **diabatic** process (from the Greek "dia" (through) and "batos" (going), as the heat goes through the envelope of the parcel).

(3) We can also quickly change the volume and pressure of the air parcel so that no heat exchange between the air parcel and the environment is possible. Such a process is called **adiabatic**. Since heat cannot be transferred in or out of the air parcel, the temperature adjustment is internal. In particular, as we will see in Chapter 6, the temperature of a *rising* air parcel will quickly drop as the pressure of the air parcel adjusts to the decreasing atmospheric pressure of the surrounding air. If the temperature of the air parcel drops to the dew point, saturation is reached and a **cloud** forms.

Summary

Water can exist in three states, or **phases**, in the atmosphere: solid (ice crystals, snow, hail), liquid (cloud droplets, raindrops), and gas (water vapor).

Water *absorbs* **latent heat** to change phase from solid to liquid to gas, which cools the surrounding air in the process. Conversely, water *releases* latent heat when changing from the gas phase to liquid to solid, which warms the surrounding air in the process.

A volume of air is said to be **saturated** when it contains the maximum amount of water vapor that can exist in equilibrium. Thus, it is useful to define the **water vapor pressure** (e), describing the amount of water vapor present in the air, and the **saturation water vapor pressure** (e_s), describing the *maximum* amount of water vapor that the air can contain.

The **relative humidity** (RH) is defined as $e/e_s \times 100$ and tells us how far (low RH) or close (high RH) the atmosphere is to saturation.

The saturation water vapor pressure depends only on temperature, and increases exponentially with temperature as described by the Clausius–Clapeyron formula.

The **dew point temperature** (T_d) is the temperature at which saturation occurs when the air is cooled. If the temperature (T) decreases to the dew point (T_d), condensation takes place.

Dew forms when condensation takes place on blades of grass and similar features near the ground. If the dew point is below $0\,°C$, then we observe the formation of **frost**. If condensation takes place on **aerosols** above the surface, then we observe the formation of cloud droplets (cloud in the form of **fog**).

The difference between the temperature and the dew point temperature is called the **dew point depression**. When the dew point depression is small (temperature and dew point are close), the air is relatively moist. When the dew point depression is large (temperature and dew point are far apart), the air is relatively dry.

The dew point can be used to identify humidity fronts ("dry lines") in weather maps, the passage of fronts in meteograms, and moist/dry layers of the atmosphere in radiosonde profiles.

Since the temperature (T) indicates the saturation water vapor pressure (e_s) and the dew point temperature (T_d) indicates the actual water vapor pressure (e), we can use T and T_d to calculate the relative humidity (RH).

An air parcel can become saturated due to: (1) evaporation of water into the parcel at constant pressure and temperature, increasing the vapor pressure to the saturation value; (2) a decrease in temperature to the dew point by loss of heat from the parcel to the environment at constant pressure; or (3) a decrease in temperature by adiabatic expansion as the parcel is brought to lower pressure through lifting.

From Fog to Clouds

Saturation, condensation, relative humidity, and dew point temperature provide us with all we need to understand how water vapor condenses to form clouds.

CHAPTER 6

Cloud Formation

We understand the basic mechanism behind condensation and the formation of dew: decrease the temperature until saturation is reached, and water vapor will start leaving the gas phase to form dew or frost. But what does it take for a cloud to form in the atmosphere? Why would the temperature decrease? What does water vapor condense upon in the absence of blades of grass and solid objects? In this chapter we will describe the cloud formation process and the conditions that are conducive to the formation of different types of clouds.

Clouds are made of water droplets (in addition to ice crystals, which we will describe in Chapter 7). These water droplets necessarily form by condensation of water vapor into liquid droplets. We mentioned in Chapter 5 that such droplets could, in theory, form by **homogeneous nucleation**: water molecules condensing onto each other, in supersaturated conditions, i.e., under very high relative humidity, about 400%. But, in practice, cloud droplets always form by **heterogeneous nucleation**. The water molecules condense onto a solid surface found amongst the many **aerosols** available in the atmosphere: solid particles such as salt, dust, pollen, and pollutants from vehicles and factories. Water vapor uses these aerosols, also called **cloud condensation nuclei**, as anchor points, from where further condensation can take place at 100% relative humidity.

Box 6.1.   Fog

Figure 6.1.1.   Evaporation fog.

Figure 6.1.2.   Radiation fog.

Figure 6.1.3.   Advection fog.

Fog is a special case in the cloud family: it is a cloud that touches the surface, and therefore does not require lifting and adiabatic cooling. Recall that we can saturate an air parcel by either adding water vapor or decreasing its temperature (Section 5.7.5), which can be accomplished in several ways.

(1) When cold air is in the presence of a warmer water surface, such as the Gulf Stream in the North Atlantic Ocean or a warm lake at the end of summer, water evaporates into the colder air and quickly saturates it, forming water droplets and fog. Recall that the saturation water vapor pressure is lower at colder temperatures (Figure 5.4) and little water vapor is required to saturate the air. This type of fog, usually called **evaporation fog** (Figure 6.1.1), corresponds to the first situation described in Section 5.7.5, where e increases to become equal to e_s.

(2) More often, however, warmer air is in the presence of a cold surface, in which case heat is transferred from the warm air to the underlying surface and the air temperature decreases. If the temperature drops to the dew point, water droplets form, resulting in fog. This typically happens after a clear night, when the ground has been cooling by emission of infrared radiation. The overlying air cools down by conduction and a temperature inversion forms close to the ground (see Figure 4.14(b)). That is why we often refer to this type of fog as **radiation fog** (Figure 6.1.2).

(3) However, warm air can also be in the presence of a colder surface because it *moves over* colder land or water. Horizontal motion is called "advection," and we refer to this type of fog as **advection fog** (Figure 6.1.3). For example, when relatively warm marine air from the Pacific Ocean moves eastward over the colder waters bordering the coast of California, it cools to the dew point and fog forms. As it continues eastward, it invades San Francisco Bay and may linger there for hours or days. A similar situation can occur over the cold Labrador Current in the North Atlantic Ocean.

Advection fog also forms when warm air moves over a colder land surface, as is typical in southern England in winter, when relatively warmer marine air that spent some time over the English Channel invades the southernmost part of the island and cools when in contact with the cold surface. The fog in southern England is notorious for reducing visibility and disrupting both road and air traffic. In London, which had a long history of burning coal in the nineteenth and twentieth centuries, fog would often form on particles of smoke, creating a thick and sickening mixture of smoke and fog that came to be called **smog**.

While aerosols provide a seed for the formation of water droplets, we still need a mechanism for bringing the air to saturation. We concluded Chapter 5 by describing three ways by which we can saturate an air parcel: by adding water vapor, decreasing the temperature by removing heat, or decreasing the temperature internally through an adiabatic process. The first two processes are effective close to the surface, and lead to the formation of **fog** (Box 6.1). Higher up in the atmosphere, although radiative cooling of moist air is an effective cooling mechanism, especially in the upper troposphere, clouds mostly form through the third process. So we focus now on adiabatic processes, and how they lead to different types of clouds.

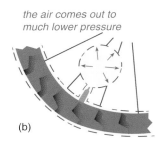

Figure 6.1. Adiabatic cooling. (a) At first, the air is in the bike tire at high pressure. (b) When it rushes out, it expands quickly without mixing or exchanging heat with the surrounding air. The blue envelope with dashed lines indicates that the air inside the tire in (a) can be viewed as an isolated air parcel that expands outside of the tire, increasing its volume and decreasing its pressure and temperature.

6.1 Adiabatic Processes

Before we investigate how air can cool to saturation, it proves useful to consider the opposite process, where we *warm* an air parcel. We are already familiar with a simple way to warm the atmosphere: heat the surface by absorption of solar energy, and transfer the heat to the lower atmosphere by conduction. The absorbed heat will be quickly transferred upward by convection and mixing.

When an air parcel is heated by conduction, energy is input directly into the parcel without actively changing its pressure or volume. We could, however, obtain a similar result by compressing the air, for example by quickly changing the pressure and volume of the air parcel (recall Box 1.3 and the ideal gas law). That is what happens when we compress air in a bike pump. You might have noticed that the pump itself warms to the touch, as the air inside heats up. The temperature increases because we exert *work* on the air. We contribute our own energy (our muscle work) to the air as we impact the air molecules with the pump piston. As the energy imparted by our muscle work is converted to heat, the total energy is conserved, in agreement with the law of conservation of energy (recall Box 1.1). But we must now be more specific about conservation of energy, which takes the form of the first law of thermodynamics – see Box 6.2.

In both cases (conduction and compression), we have increased the heat content of the air parcel, and therefore its temperature. By inverse reasoning,

we can assume that the heat content of the air can decrease: (1) by conduction of heat from inside the parcel to the outside environment, which is what happens at night when the lower atmosphere cools by conduction of heat to the surface; or (2) when work is exerted by the air such that it loses energy. A good example of this second situation is the reverse operation of pumping air into a tire: when the air rushes out of the bike tire after the valve is opened. Since air in the tire is at higher pressure than the air outside the tire, air rushes out of the tire and expands to occupy a larger volume than it did in the bike tube alone. As it does, it must push the surrounding air aside, and this requires work (Figure 6.1). In other words, the air inside the tire exerts work on the surrounding air and, in doing so, loses energy, which cools the air. In fact, if you place your hand in front of the valve, you will feel that the air is colder than the air you pumped in. Work has an energy cost, and it is paid out of the heat content of the air doing the work.

These temperature changes are possible because the two processes (compression in the pump and escape out of the tire) are happening very quickly – so quickly that no exchange of energy is possible between the air being compressed or released and the environment (i.e., the surrounding air). To help us visualize this abstract concept from thermodynamics, we can imagine that the air inside the tire is like an isolated air

Box 6.2.   The First Law of Thermodynamics

In the nineteenth century, several scientists formulated what is now called the first law of thermodynamics, which says that an increment of heat added to a system can change the temperature of the system or be used by the system to perform work. Formally,

$$\Delta Q = c_p \Delta T + W$$

where Q stands for heat, T for temperature, W for work, c_p for specific heat of the substance being heated, and Δ indicates a change.

In an adiabatic process, there is no exchange of heat between the system and its surroundings. Therefore, the total amount of heat in the system does not change and $\Delta Q = 0$. Mathematically, this means that

$$c_p \Delta T + W = 0$$

and

$$c_p \Delta T = -W$$

In other words, work exerted on the system ($W < 0$) or by the system ($W > 0$) must be compensated by a change in temperature. In the bike pump example presented in the text, the work exerted by our muscles is positive for us, and therefore negative for the air (it is *received* by the air), therefore $-W$ is positive and ΔT is positive: the air warms (adiabatic warming).

In the bike tire example, work is exerted by the air rushing out of the tire and pushing against its surroundings, therefore W is positive and ΔT is negative: the air cools (adiabatic cooling).

parcel (Figure 6.1(a)). In the first case (compression), as we push on the piston, the volume of this closed system decreases very quickly, and the pressure increases, along with temperature. In the second case (air escape from the tire), the air in the tire mostly does not mix with the surrounding air when it escapes, but remains inside an imaginary envelope that expands out of the tire (Figure 6.1(b)). It remains mostly a closed system because the expansion is very fast. The volume of this system increases as the pressure decreases, along with temperature.

A process in which no energy is being exchanged between the system and its surroundings is called an **adiabatic process**. The compression in the bike pump causes **adiabatic warming**, and the expansion of air outside the tire causes **adiabatic cooling**. A more formal explanation is provided in Box 6.2.

6.2 Adiabatic Processes in the Atmosphere

How is all this information about adiabatic processes relevant to cloud formation? The reason is that air parcels undergo adiabatic expansion whenever they move upward. Recall that atmospheric pressure decreases rapidly with height (Figure 1.6). For example, consider an air parcel at the surface, which you can imagine as a balloon of air with an invisible boundary. As it rises, the balloon quickly finds itself in an environment where atmospheric pressure is much less than inside the balloon, which makes the balloon expand outward against the surrounding air at lower pressure. Because the pressure adjustment is instantaneous and adiabatic (no exchange of air, no exchange of heat), the air molecules in the air parcel perform work on the surrounding environment and lose energy as a result: the air parcel cools. The reason the air temperature inside the parcel does not adjust to the outside temperature is because the heat exchange processes, such as mixing and radiation, are much slower than adiabatic expansion, and as a result the parcel cools.

Now you probably recognize a familiar scenario: if we lift the air parcel high enough, its temperature will decrease sufficiently so that the saturation water vapor pressure becomes equal to the actual water vapor content: the air in the parcel will have reached saturation and condensation will take place. We are on our way to making a cloud.

How far does an air parcel need to rise before it reaches saturation? Should we always observe the formation of a cloud when air rises? What are the environmental factors that control the formation of clouds? Before we answer these questions and continue our exploration of cloud formation, let us have a quantitative look at adiabatic expansion.

6.3 Dry Adiabatic Lapse Rate

The vertical structure of the atmosphere is relatively well known. It is known on average (see Figures 1.6 and 3.2, for example), and it can be known in detail on any given day by releasing a radiosonde, which makes measurements on its way up through the atmosphere. We mentioned in Chapter 3 that the **environmental lapse rate** (the rate at which temperature decreases with height) is about 6.5 °C/km on average, but that it can be very different on any given day and at any given location. The environmental lapse rate is set up by a mix of radiative and convective processes, as well as horizontal air motion (i.e., advection).

Since we know the rate at which pressure decreases with height, we can calculate the work that the air parcel does as it is lifted above the surface, and therefore the rate of temperature decrease (see Box 6.2). The rate of temperature decrease due to adiabatic lifting and expansion is called the **dry adiabatic lapse rate** and is about 10 °C/km (or 5.5 °F/1000 ft) for *unsaturated* air. (We will deal with moisture and saturation shortly.)

Note, and this is an important point, that the concept of an adiabatic lapse rate applies to air *inside* the parcel, which is different from the concept of an environmental lapse rate *outside* the parcel. When establishing an environmental temperature profile and calculating an environmental lapse rate, we are simply measuring temperature at each altitude with a thermometer. Keep in mind that this vertical temperature structure is determined by several processes, including radiative and convective processes, as well as advection. When calculating the dry adiabatic lapse rate, however, all that matters is that the air parcel is rising and losing energy as it works against the surrounding environment during expansion (Figure 6.2).

Note further that, by adiabatic expansion, the air parcel cools at a much faster rate than the surrounding air outside the parcel because that air is not rising.

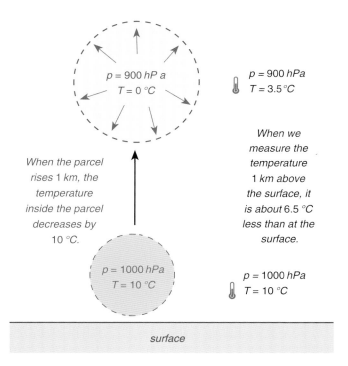

Figure 6.2. The temperature inside a rising air parcel decreases at the dry adiabatic lapse rate (10 °C/km) while the temperature measured by a thermometer at different altitudes decreases at the environmental lapse rate (6.5 °C/km on average).

And because there is no exchange of heat between the parcel and its surroundings, the air parcel usually becomes colder than the surrounding air. Therefore, the temperature drops in the air parcel independent of the temperature of the surrounding air. This difference in temperature between the air parcel and the surrounding air is critical to understanding cloud formation, and will occupy us for much of the rest of this chapter.

Finally, note that the adiabatic lapse rate applies in the downward direction as well: when an air parcel subsides (another word for going down) by 1 km, it warms by 10 °C. That is because, finding itself in an environment where the pressure is greater, the air inside the parcel is being compressed. The surrounding air now performs work on the parcel, which results in an increase in temperature (Figure 6.3). This is exactly the same process as the bike pump example, where the pump warms as we increase the pressure in the tire.

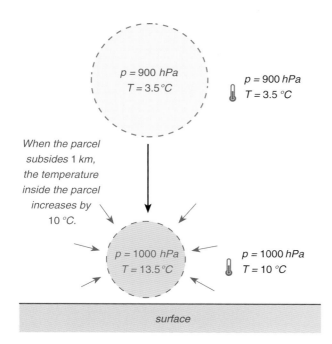

Figure 6.3. The temperature inside a subsiding air parcel increases at the dry adiabatic lapse rate (10 °C/km) due to adiabatic compression.

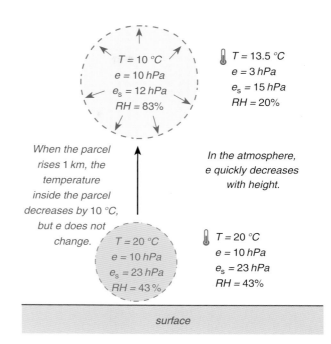

Figure 6.4. The relative humidity (RH) inside a rising air parcel increases due to adiabatic cooling and the decrease in saturation water vapor pressure.

6.4 Relative Humidity

When an air parcel rises, its water content does not change, since there is no exchange of air, and therefore no exchange of moisture with air surrounding the parcel: to a first approximation, the water vapor pressure (e) stays the same. [Note that, strictly speaking, p and e both decrease proportionally, following Dalton's law – see Box 5.1 – but, for demonstration purposes, we will assume that e remains constant.] The temperature, however, decreases due to adiabatic cooling, and we know that the saturation water vapor pressure (e_s), being a function of temperature, also decreases (Figure 5.4). As a result, the relative humidity (RH) increases. For example, Figure 6.4 shows an air parcel at 20 °C containing 10 hPa of water vapor. Using Figure 5.4 we can calculate that e_s = 23 hPa, and therefore:

$$RH = e/e_s = 10/23 = 0.43 = 43\%$$

After rising 1 km, the air parcel has cooled by 10 °C (at the dry adiabatic lapse rate) to reach a temperature

$T = 10\ °C$, which corresponds to $e_s = 12\ hPa$, so the new relative humidity is

$$RH = e/e_s = 10/12 = 0.83 = 83\%$$

As the temperature decreases, the relative humidity increases (consistently with what we found in Chapter 5 for the formation of dew). Note again that RH does not increase because there is more moisture in the air parcel – only because the capacity decreases, and there is more moisture *in relation* to the maximum capacity.

Again, since the temperature decrease applies to the parcel, and not the air surrounding the parcel, the increase in relative humidity pertains only to the air in the air parcel. The water vapor content of the environment at that altitude (outside the parcel) might be only $e = 3\ hPa$, because most of the water vapor is concentrated near the surface and e quickly decreases with height. If the surrounding air temperature is 13.5 °C, which corresponds to $e_s = 15\ hPa$, the relative humidity of the surrounding air is:

$$RH = e/e_s = 3/15 = 0.20 = 20\%$$

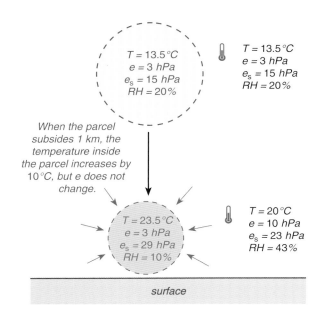

Figure 6.5. The relative humidity (RH) inside a subsiding air parcel decreases due to adiabatic warming and the increase in saturation water vapor pressure.

The relative humidity is much lower outside than inside the parcel. That is because the air parcel has been lifted with its water vapor, and is therefore necessarily more moist than its surroundings, where there is very little moisture.

The reverse is also true: when an air parcel subsides in the atmosphere, its water vapor content does not change, but its temperature increases due to compression and adiabatic warming, which increases its saturation water vapor pressure (Figure 6.5). As a result, the relative humidity decreases. We will see that these circumstances are typical of anticyclones, as well as the downwind side of mountains.

6.5 Moist Adiabatic Lapse Rate

What happens if an air parcel rises and cools sufficiently that it reaches saturation, but we keep lifting it? We know that some water vapor will leave the gas phase and condense on aerosols to form cloud droplets. We also know that the process of changing from the gas to liquid phase releases latent heat (the additional energy that was allowing the water molecules to be at a higher level of molecular activity). This latent heat warms the air

Box 6.3. The First Law of Thermodynamics, Revisited

When a rising air parcel is already saturated, some water vapor condenses to form cloud droplets and latent heat is released. Since this latent heat is released in the parcel and warms the air inside the parcel, the process is not adiabatic: the change in heat content of the parcel, ΔQ, is positive. Formally,

$$\Delta Q = c_p \Delta T + W > 0$$

and

$$c_p \Delta T = -W + \Delta Q$$

The change in temperature of the parcel is due both to work exerted by the system and to the release of latent heat into the system. Because W is greater in magnitude than ΔQ, the change in temperature is still negative, but it is less than what it would be without the release of latent heat.

parcel (see Box 6.3 for a more formal approach). Thus, we are dealing with two competing effects: adiabatic cooling due to expansion, and latent heating due to the condensation of water. As a result, a saturated air parcel rising 1 km in the atmosphere cools by about 10 °C due to adiabatic cooling, but also warms by about 4 °C due to latent heating (Figure 6.6). The net result is that the air parcel cools by only 6 °C/km. We call this the **moist**, or **saturated**, **adiabatic lapse rate**. (In fact, this rate depends strongly on temperature, because it depends on the saturation vapor pressure, but we will use this value since it applies broadly.)

Note that, in our example, the air parcel contained 19 hPa of water vapor (e) at the surface, but now contains only 13 hPa after rising 1 km, because the saturation water vapor pressure (e_s) has dropped to 13 hPa and 6 hPa of water vapor has condensed to form cloud droplets. In other words, e follows e_s and they both decrease at the same rate as the air rises – recall that e can never be greater than e_s.

Let us apply these new concepts to a real-world situation.

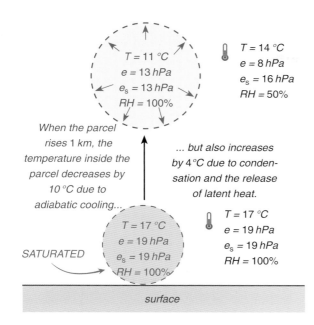

When the parcel rises 1 km, the temperature inside the parcel decreases by 10 °C due to adiabatic cooling...

... but also increases by 4 °C due to condensation and the release of latent heat.

SATURATED

surface

Figure 6.6. The temperature inside a rising parcel saturated with water vapor decreases at the moist adiabatic lapse rate (6 °C/km).

6.6 Orographic Lifting

We now have all the ingredients to make a cloud through lifting. We know that, if air rises, it expands adiabatically and cools at the dry adiabatic lapse rate, and, if lifted enough, can reach saturation. The simplest mechanism to lift air is a mountain range in the path of horizontal wind. As the air runs into the mountain range, it rises over the mountain (on the windward side) and subsides on the other side (the leeward side, also called the "lee" of the mountain). Because the air is lifted by orographic features (i.e, by the topography), we refer to this lifting mechanism as **orographic lifting**.

Let us consider an air parcel at 24 °C, containing 8 hPa of water vapor, and ascending a 5 km mountain range (Figure 6.7). We can see from Figure 5.4 that, at 24 °C, the saturation water vapor pressure of the air parcel is 30 hPa. Therefore, its relative humidity is $8/30 = 0.27 = 27\%$ at the foot of the mountain.

As the air parcel rises up the mountain slope, the pressure decreases, and the parcel temperature decreases at the dry adiabatic lapse rate due to adiabatic expansion: it decreases by 10 °C for each kilometer gained in elevation. At any elevation, we can

calculate the new temperature, the saturation water vapor pressure, and therefore the relative humidity, as summarized in Table 6.1.

We can see that saturation is reached at an elevation of 2 km, where the temperature has dropped by 20 °C, and e_s has dropped to 8 hPa, becoming equal to the actual water vapor pressure (RH = 100%). This elevation marks the altitude of the cloud base and is called the **Lifting Condensation Level** (LCL). With additional lifting, the air parcel temperature keeps decreasing, which means the saturation vapor pressure e_s decreases, and the excess water vapor condenses out of the vapor phase onto condensation nuclei to form cloud droplets: we are now inside the cloud.

Since water vapor condenses (above 2 km), latent heat is released and the air parcel now cools at the *moist* adiabatic lapse rate (6 °C/km). As water leaves the gas phase, the actual water vapor content of the air parcel decreases – it "follows" the saturation water vapor pressure, down to 2 hPa at the top of the mountain.

Where does this condensed water go? We have assumed that air parcels exchange no mass or energy with the surrounding air, but we have to make an exception once clouds form, since precipitation processes can cause the water to leave the air parcel permanently (for example, as rain or snow reaching the ground).

What happens as the air parcel subsides back down the other side of the mountain range? As soon as it starts subsiding, it is compressed adiabatically and the temperature increases. With the slightest temperature increase, the saturation water vapor pressure also increases, and e_s becomes greater than e: the relative humidity drops below 100% and the air parcel is no longer saturated when it subsides. Therefore, the air parcel temperature increases at the *dry* adiabatic lapse rate all along the lee slope (Figure 6.8), i.e., much more than it decreased on the windward side. The whole descent is summarized in Table 6.2, where the air parcel can be seen to warm to 36 °C, which is 12 °C greater than its original temperature at the base of the mountain on the windward side. Moreover, because the air parcel temperature is high (and therefore so is the saturation water vapor pressure), and because the actual

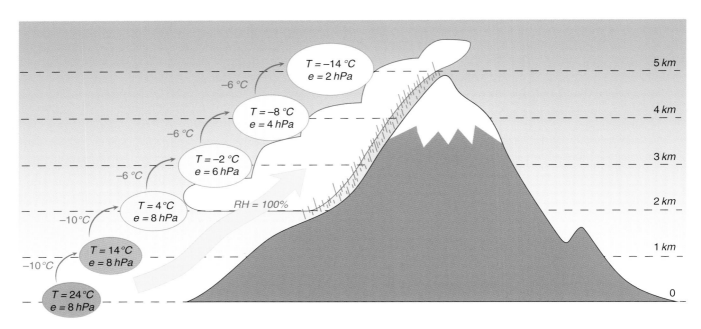

Figure 6.7. Schematic representation of an air parcel ascending a mountain, cooling due to adiabatic expansion, and reaching saturation at an altitude of 2 km (cloud base).

Table 6.1. *Example of changes in T, e$_s$, e, and RH as an air parcel rises along a mountain slope.*

Altitude (km)	Temperature (°C)	Saturation water vapor pressure (hPa)	Water vapor pressure (hPa)	Relative humidity (%)
0	24	30	8	27
1	14	16	8	50
2	4	8	8	100
3	−2	6	6	100
4	−8	4	4	100
5	−14	2	2	100

water vapor content is so low (because the water left the air parcel in clouds and precipitation on the windward side), the relative humidity is extremely low. In other words, not only is the air very warm, but also it is very dry.

These calculations are useful to explain many observations of orographic precipitation and subsidence around the world. All along the West Coast of North America, for example, the prevailing westerly winds are lifted by the coastal mountain ranges and produce enhanced precipitation, and therefore lush vegetation cover on the western slopes. The eastern slopes are comparatively warmer and drier, and are often a desert landscape. That is where the air subsides and warms by adiabatic compression at the dry adiabatic lapse rate. This is illustrated in Figure 6.9 for Washington State, where the westerlies first ascend the Coastal Range and the Olympic Mountains, then the Cascades, and finally the Rocky Mountains. Note on the satellite image (Figure 6.9(a)) how the

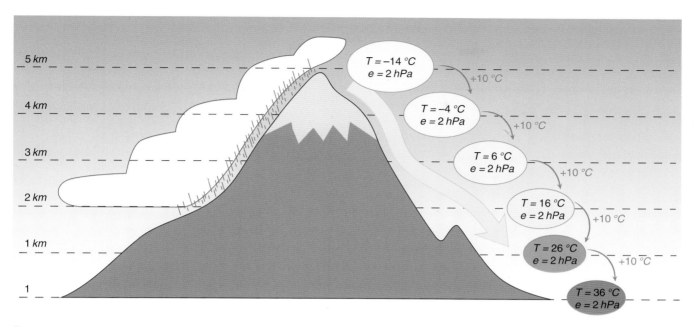

Figure 6.8. Schematic representation of the air parcel subsiding on the other side of the mountain, and warming due to adiabatic compression.

Table 6.2. *Example of changes in T, e_s, e, and RH as the air parcel subsides back down on the lee side.*

Altitude (km)	Temperature (°C)	Saturation water vapor pressure (hPa)	Water vapor pressure (hPa)	Relative humidity (%)
5	−14	2	2	100
4	−4	5	2	40
3	6	10	2	20
2	16	18	2	11
1	26	34	2	6
0	36	60	2	3

western slopes are relatively green, while Eastern Washington, in the lee of the Cascade mountains, has the brown appearance of a desert. The same picture applies to the Sierras of California and the deserts that exist in their lee. Note also that air parcels do not always subside back to sea level, but more often to an elevated plateau, so that the increase in temperature is usually not as great as described in our idealized example.

The warm and dry winds caused by subsidence on the lee side of mountain ranges are well known to people living in these locations. The **chinook**, for example, results from subsidence on the eastern slopes of the Rockies that blows into the Great Plains. The **foehn** results from Mediterranean air blowing over the Alps and descending into the plains and valleys of Central Europe. The terms "chinook" and "foehn" are used more generally to refer to winds that result

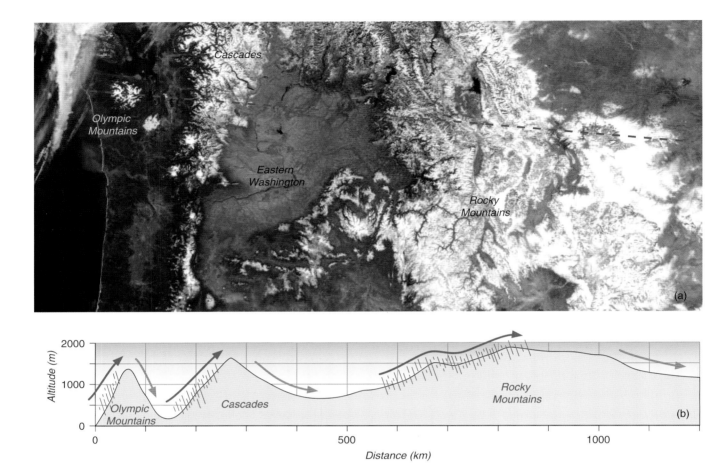

Figure 6.9. Example of orographic precipitation on the windward slopes of the Coastal Mountain Range, the Cascades, and the Rocky Mountains in the Pacific Northwest. (a) Vegetation cover and snow as captured by MODIS on the NASA Terra satellite. (b) Land elevation along the red dashed line in (a).

from orographic lifting and subsidence. Orographic precipitation and lush vegetation contrasting with warm and dry lee slopes can be observed in many other places, such as New Zealand, the southern part of South America, in Chile and Argentina, and Hawaii (Figure 6.10).

To conclude, let us emphasize that all our deductions were made with very little information: merely the temperature (T) and water vapor content (e) at the base of the mountain. Everything beyond that was based on knowledge of the adiabatic lapse rates and the mechanism of condensation. It is noteworthy that we can predict the altitude of the cloud base and the conditions in the lee of mountains with no information about the atmosphere above the surface.

Finally, let us summarize and emphasize that cloud formation requires cooling air to saturation, and that, in most cases, the cooling occurs as a result of lifting and adiabatic expansion. As for subsidence, we established that, as soon as the air subsides, it warms due to adiabatic compression. Necessarily, the saturation water vapor pressure (e_s) increases and the relative humidity drops. Consequently, clouds cannot form in sinking air without a source of moisture. In other words, *subsidence inhibits cloud formation.* We will make use of this fact shortly. In the meantime, let us explore additional mechanisms leading to the formation of clouds.

Figure 6.10. The northeastern slopes of Hawaii, exposed to the trade winds (blue arrow), are lush and green, while the southwestern slopes are in the lee of the dominant winds, and therefore much drier.

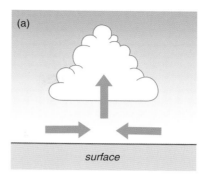

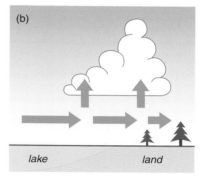

Figure 6.11. Two examples of lifting by convergence. (a) Winds converge from different directions. (b) Wind speed decreases as the roughness of the underlying surface increases.

6.7 Lifting by Convergence

When air flows from different directions and meets in the same location, it is said to **converge**, much like cars on different roadways converging on an intersection. Although in principle the air density could increase at the point of convergence, it turns out that air is more likely to move in a different direction rather than change density. When convergence happens at the ground, the only option is upward, in which case adiabatic cooling can occur, along with cloud formation and possibly precipitation (Figure 6.11(a)).

Convergence can also take place when the wind slows down along its path, while keeping a higher speed upstream (Figure. 6.11(b)). Although this is more difficult to visualize, one can see how air would tend to accumulate and pile up because the air ahead is not flowing fast enough. This is similar to a car slowing down on the highway, and forcing the cars behind it to get closer and more compact. While cars have no choice but to slow down as well, air can continue upward, possibly causing condensation and cloud formation if saturation is reached due to adiabatic cooling. This slowing effect can happen when wind blowing over a water surface, such as a lake, reaches the shore and continues to flow over land. While

water is relatively smooth and allows the wind to blow faster, land is rougher, and friction slows the wind. In such a situation, clouds are observed just downstream of the shore, over land. In winter, these clouds can produce snow.

6.8 Frontal Lifting

You might recall that a cold front occurs where a cold air mass advances against a warm air mass. We introduced a bird's eye view in Chapter 2, here reproduced in Figure 6.12(a). If we switch to a cross-section (Figure 6.12(b)), we notice that cold air, being denser, tends to sink close to the surface and to wedge underneath the warm air mass. This causes the warm air to rise, which can lead to condensation and the formation of clouds. Because this ascent of the warm air is relatively abrupt, clouds of the cumulus family tend to form, especially if the atmosphere ahead of the cold front is

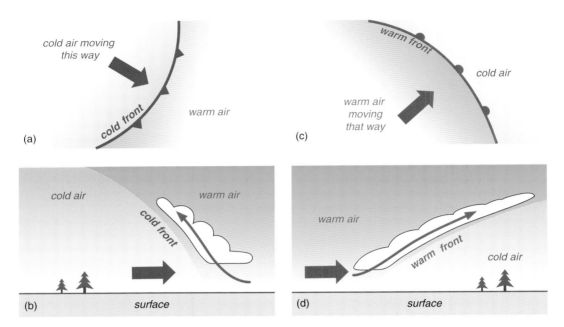

Figure 6.12. Cloud formation by frontal lifting. (a) Bird's eye view showing a cold air mass advancing against a warm air mass. (b) Cross-section showing the same cold air mass wedging underneath the warm air and lifting it up, causing the formation of cumuliform clouds. (c) Bird's eye view of a warm air mass advancing toward a cold air mass. (d) Cross-section showing the warm air overrunning the cold air and reaching saturation, causing the formation of stratiform clouds.

unstable – see Section 6.9. Cumulus clouds have more vertical development, and a cauliflower appearance that makes them easily recognizable (see Appendix 6.1).

In contrast, when a warm air mass advances against a cold air mass (Figure 6.12(c)), it does not wedge underneath, but slides upward and over the cold air mass, a phenomenon called **overrunning** (Figure 6.12(d)). As usual, clouds form as soon as saturation is reached, but since overrunning is the result of a more gentle uplift than the wedging of the cold air mass described earlier, clouds of the stratus family tend to form (stratus, altostratus, and cirrostratus). Stratus clouds have a more horizontal development and take on the form of extensive sheets of uniform clouds (see Figures 6.32 and 6.33 in Appendix 6.1).

6.9 Convection

Last but not least, clouds can form when warm air is lifted by convection – in other words, by buoyancy forces, as described by Archimedes' principle (see Box 6.4). This lifting mechanism is the most difficult to understand, but results in spectacular weather in the form of thunderstorms, so we will devote the rest of this chapter to understanding how it works and how, in general, buoyancy affects the shape of the clouds that we see in the atmosphere.

6.9.1 Stable Air

Let us consider an atmosphere described by the temperature profile shown in Fig. 6.13, which could be measured by a radiosonde. (By default, we use red to show the environmental lapse rate – red like the color of a thermometer). It starts with a temperature of $T_e = 20\,°C$ at the surface, and quickly drops with height. Let us now imagine an air parcel at sea level that is lifted from the surface. At first, the air parcel temperature is equal to that of the environment ($T_p = 20\,°C$). As soon as the air parcel is lifted, its temperature decreases at the dry adiabatic lapse rate, $10\,°C/$ km (let us not worry about humidity right now). If it

Box 6.4.   Buoyancy and Archimedes' Principle

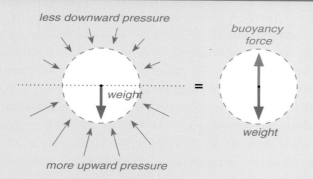

Figure 6.4.1.   Force balance between the weight of an air parcel and the buoyancy force applied to that parcel in the case of equilibrium.

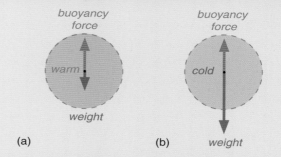

Figure 6.4.2.   (a) Warm air rises. (b) Cold air sinks.

Any object bathed in the atmosphere is experiencing pressure forces from surrounding air molecules. Since atmospheric pressure decreases with height, the downward force exerted by air molecules above the object is less than the upward force exerted by air molecules below. This pressure differential results in a net force, called the *buoyancy force*, that pushes the object upward. When the weight of the object is exactly balanced by the buoyancy force, the object is in equilibrium – it is floating (Figure 6.4.1). This would be the case of a balloon floating at a given height above ground, or a submarine stable at a given depth in the ocean (the same reasoning applies in water).

Another way to look at this balance of forces is to see that the buoyancy force is equal in magnitude to the weight of the floating balloon – and opposite in direction. If it were not so, the balloon would either be pushed up (if the buoyancy force were greater than the weight) or down (if the weight were greater than the buoyancy force).

Now for a thought experiment: replace the balloon with the parcel of air that *would be* there if the balloon were not – in other words, replace the balloon with the air that was *displaced* by the balloon. That air parcel would have the same volume as the balloon, and, since it would also be floating and stable, it must have the same weight as the balloon. Therefore, the balloon and the displaced air are interchangeable, and the buoyancy force, which is equal in magnitude to the weight of the balloon, is also equal to the weight of the displaced air. This is, in essence, Archimedes' principle: "When an object is immersed in a fluid, it is subject to an upward buoyancy force equal in magnitude to the weight of the displaced fluid."

Let us go one step further. Replace the floating balloon with a parcel of identical volume, but containing *warm* air. Recall that warm air has a lower density, and therefore (the volume being equal) the warm parcel contains less mass of air than a cold parcel. As a result, the weight of the warm parcel is less than the weight of the earlier parcel at environmental temperature. In particular, it is not sufficient to balance the upward buoyancy force. Therefore, there is a net upward force that pushes the parcel up: warm air rises Figure 6.4.2(a)). Replace the parcel with cold, dense air, and the weight will be increased and will overcome the buoyancy force. The net force will push the air parcel downward: cold air sinks Figure 6.4.2(b)).

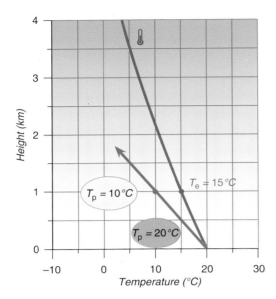

Figure 6.13. Example of a stable temperature profile (in red). The blue arrow represents the decrease in temperature inside an air parcel as it is lifted from the surface.

rose 1 km, its new temperature would be $T_p = 10\,°C$ (see the blue arrow in Figure 6.13), which we can compare to the temperature of the environment at that height, outside the air parcel. The environmental temperature (as indicated by the red curve) is $T_e = 15\,°C$. Therefore, the air parcel would be *colder* than its surroundings, more dense, and would tend to sink back down. In these circumstances, air parcels return to where they came from. We say that the atmosphere is **stable**, or that we have a stable temperature profile. (See Box 6.5 for a complete description of stability.)

6.9.2 Unstable Air and Thermals

By contrast, consider the atmosphere described by the temperature profile in Figure 6.14, where temperatures close to the surface are higher than in the previous example, presumably due to the absorption of solar radiation by the surface and transfer of heat to the lower atmosphere by conduction. Let us repeat the previous exercise and lift an air parcel 1 km above the surface, starting at a temperature of $30\,°C$.

Its new temperature is $T_p = 20\,°C$, which is *greater* than the environmental temperature at that height ($T_e = 13\,°C$). Therefore, even though it has risen and cooled, it is still *warmer* than its surroundings. This means that the air parcel is less dense than its surroundings, and therefore still subject to an upward buoyancy push. We say that the atmosphere (or at least that particular layer of the atmosphere) is **unstable**.

As the air parcel continues to rise, its temperature continues to decrease at the same rate, until it becomes colder than its surroundings, in which case it stops rising because its density matches that of the surroundings and the buoyancy force becomes zero. By following the blue arrow in Figure 6.14(a), we can see that this happens above 3 km, where the temperature profile is once again stable.

On a sunny afternoon, many such warm parcels might experience such an upward buoyancy push. We cannot imagine, however, an entire layer of warm air close to the surface rising all at the same time. Rising air needs to be replaced by some fresh air coming from another direction – the void created by the rising air needs to be filled, in a sense. In practice, the atmosphere organizes itself in such a way that warm air rises in places, while cooler air from aloft sinks in between, as shown in Figure 6.15. The pockets of rising warm air, which provide a natural upward push, are called **thermals**. They are invisible to the naked eye, but are perceptible to the senses of birds and glider pilots, who learn to fly from one thermal to another, and can remain airborne indefinitely.

6.9.3 Stable vs. Unstable

If you pay attention to the environmental temperature profiles in Figures 6.13 and 6.14, you will notice that the dry adiabatic lapse rate, the blue line, does not change (of course). What determines the stability of the atmosphere is the particular tilt of the *environmental* temperature profile, of the red curve – in other words, the environmental lapse rate. Whenever the environmental temperature profile is leaning to the left of the adiabatic lapse rate, the atmosphere is unstable (because, again, a rising air parcel will necessarily remain warmer than the environment). Whenever the environmental temperature profile is leaning to the

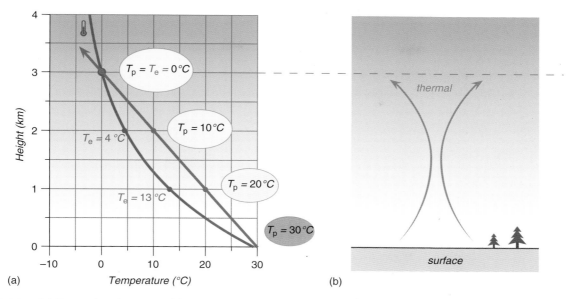

Figure 6.14. (a) Example of an unstable temperature profile (in red). The blue arrow represents the decrease in temperature inside an air parcel as it is lifted from the surface. The air parcel is buoyant up to about 3 km, where it stops rising. (b) Resulting thermal.

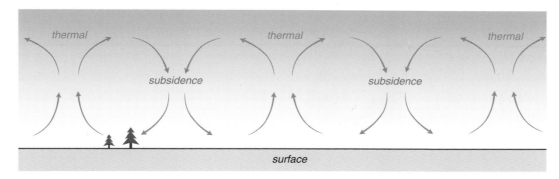

Figure 6.15. Organization of convective cells with alternating thermals and regions of subsidence.

right of the blue line, the atmosphere is stable, as summarized in Figure 6.16. If the environmental temperature profile is exactly aligned with the adiabatic lapse rate, it is neither stable nor unstable: we say that it is **neutral**, or **neutrally stable**. You can verify for yourself that, with a neutral temperature profile, a lifted air parcel will neither continue to rise nor sink back down. Being at the same temperature as its surroundings, it simply stays at its new position (see also Box 6.5).

Certain situations tend to make the atmosphere unstable, e.g., large-scale lifting and cooling aloft. Conversely, subsidence warming, for example, tends to make the atmosphere more stable. An interesting case occurs when the environmental temperature profile is increasing with height (a temperature inversion), as is the case after a clear night, after the lower atmosphere has cooled by conduction of heat to a cold surface (Figure 6.17). Since the atmosphere itself becomes warmer with height, and a rising air parcel can only cool when lifted, any lifted air will necessarily experience a strong buoyancy force to sink back down. That is why temperature inversions are notorious for *trapping* air at low elevations, preventing the dissipation of fog, for example, or trapping pollution and causing health hazards (discussed in more detail in Chapter 14).

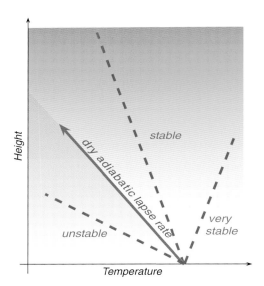

Figure 6.16. Various environmental lapse rates (red) in relation to the dry adiabatic lapse rate (blue). Temperature profiles in the blue sector tend to be unstable, while they tend to be stable in the red sector.

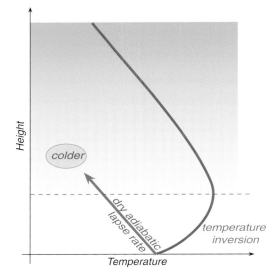

Figure 6.17. A temperature inversion is necessarily stable, as a lifted air parcel necessarily cools down, while the environmental temperature increases with height.

6.9.4 Fair-Weather Cumulus Clouds

Let's add water to our understanding of stability to see how cloud formation fits in the picture. For example, consider the temperature profile shown in Figure 6.18 and lift a parcel from the surface. The air parcel has a temperature $T_p = 30\,°C$ and contains $e = 12\,hPa$ of water vapor. Using Figure 5.4, we can calculate that the saturation water vapor pressure at $30\,°C$ is $e_s = 43\,hPa$. Therefore, the air parcel is not saturated and cools at the dry adiabatic lapse rate ($10\,°C/km$) when lifted. After rising 1 km, it has cooled to $20\,°C$, which corresponds to $e_s = 23\,hPa$. The parcel is still not saturated ($e < e_s$), and it is still warmer than the environment ($T_p > T_e$). Therefore, it keeps rising due to the upward buoyancy force. So far, we are following the *unstable* scenario described previously.

After rising one more kilometer, however, the air parcel has cooled down to $10\,°C$, at which point $e_s = 12\,hPa$. We have reached saturation ($e_s = e$, RH = 100%), and we know that any further lifting will lead to condensation and cloud formation. Recall that this critical threshold is called the **lifting condensation level** (LCL) as it is the location where the cloud base is realized.

We also know that, due to latent heat release, any cooling from the air parcel rising further will occur at the *moist* adiabatic lapse rate ($6\,°C/km$). Since the air parcel is still warmer than its surroundings ($T_p > T_e$), it keeps rising and, after one more kilometer, will cool to $4\,°C$. The saturation water vapor pressure at this new temperature is $8\,hPa$, which means that $4\,hPa$ of water vapor have condensed to form the lower portion of a cloud. As for the air parcel ascending the mountain, the water vapor pressure decreases and "tracks" the saturation water vapor pressure ($e = e_s = 8\,hPa$). The relative humidity remains 100% and the air parcel is saturated at all times.

Is the air parcel still buoyant? Its new temperature is equal to the environmental temperature ($T_p = T_e = 4\,°C$), therefore it stops rising, and the cloud does not develop much above 3 km (the cloud top). We are in the presence of a relatively small cloud. The upward motion of air associated with convection tends to form clouds of the cumulus type (bright white bundles with a cauliflower appearance and sharply defined boundaries). Since these clouds tend to form on sunny days when the surface is heated by absorption of solar radiation and the atmosphere is only mildly unstable, we often call them **fair-weather cumulus** (Figure 6.19).

Box 6.5. Stability

stable

Figure 6.5.1. Stable situation.

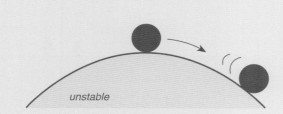

unstable

Figure 6.5.2. Unstable situation.

neutral

Figure 6.5.3. Neutrally stable situation.

conditionally unstable

Figure 6.5.4. Conditionally unstable situation.

Lifted air parcels can be either stable, unstable, or neutrally stable, depending on what happens when they are lifted. To understand better how the concept of stability applies to the atmosphere, it is useful to go back to the more general definition of stability in physics, and to use the classic example of a ball on a curved surface.

In Figure 6.5.1, the ball is at the bottom of a *valley*. If given a small push, left or right, the ball will tend to return to its original position. We say that the ball is in a **stable** position.

In Figure 6.5.2, the ball is at the top of a *mountain*. If given a small push, the ball will continue rolling away from its original position. We say that the position of the ball is **unstable**.

By analogy, a temperature profile such as shown in Figure 6.13 is described as *stable*, because a lifted parcel will tend to return to its original position, whereas a temperature profile such as described in Figure 6.14 is described as *unstable* because a lifted parcel will continue to rise, away from the surface.

In between those two situations, there is the possibility that the ball is on a flat surface, in which case, if given a small push, the ball will neither return to its original position nor continue rolling. It will simply stop at its new position (Figure 6.5.3). We say that the ball is **neutrally stable**.

A temperature profile that would exactly follow the adiabatic lapse rate would be described as neutral because a lifted air parcel would cool at the same rate as its surroundings. If lifted, the air parcel would be at the same temperature as the environment and would stop at its new position.

There exists a final type of situation in which the ball can have a stable or unstable behavior depending on the circumstances. In Figure 6.5.4, the ball will be stable if given a small push, but will become unstable if given a large enough push that it reaches one of the adjacent summits. We say that the position of the ball is **conditionally unstable**.

In the atmosphere, conditional instability occurs when lifted air parcels are stable when the lift is small, but become unstable when the upward displacement is sufficiently large (Figure 6.22).

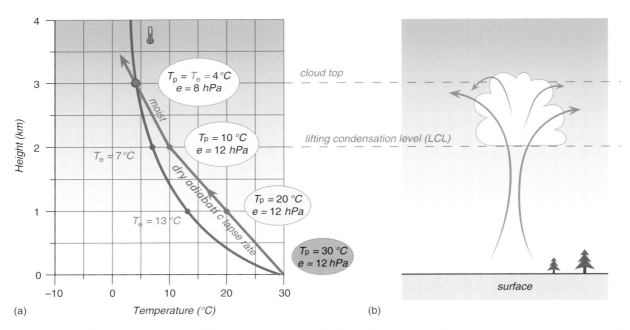

Figure 6.18.   (a) Example of an unstable temperature profile (in red) with formation of a fair-weather cumulus. The blue and green arrows represent the decrease in temperature inside an air parcel as it is lifted from the surface: first at the dry adiabatic lapse rate (blue) and then at the moist adiabatic lapse rate above 2 km (green). The air parcel is buoyant up to about 3 km, where it stops rising. (b) Resulting cumulus cloud.

Figure 6.19.   Fair-weather cumulus clouds.

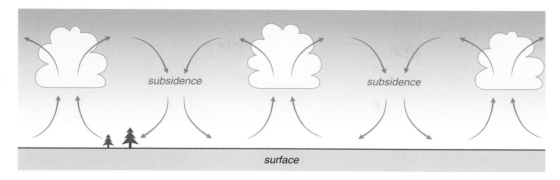

Figure 6.20. Formation of fair-weather cumulus clouds within the ascending branch of convective cells.

In the same way that thermals are organized within convective cells, fair-weather cumulus clouds tend to appear within the ascending branch of convective cells and to alternate with regions of subsidence where there is no cloud (Figure 6.20). That is why they often appear as isolated clouds distributed at somewhat regular intervals. Note also that, because the lower atmosphere is often relatively uniform in temperature and humidity, all air parcels reach saturation at about the same height (the LCL). That is why cumulus clouds often have a flat cloud base, and why all cumulus clouds present at a given time often have their cloud base at around the same height.

If you observe cumulus clouds for a few minutes, you will notice that they form and dissipate relatively quickly. Due to the turbulent motion caused by convection, they mix with the surrounding air, especially along their boundaries. This process is called **entrainment**. Since the surrounding air is drier, and not saturated, the entrainment of environmental air into the cloud tends to lower the relative humidity inside the cloud, causing evaporation and cooling of the warm buoyant air. In the process, however, moist air from the cloud is being mixed into the environment, which *moistens* the atmosphere. Moisture has been lifted from surface level up to higher elevations, increasing the relative humidity. Therefore, as cumulus clouds grow and evaporate, the temperature and humidity profiles evolve. In particular, the atmosphere can become more unstable due to moistening the air at higher levels. As the clouds grow in size and number, they start to occupy, and "congest," the sky, developing into **cumulus congestus** (Figure 6.21). If the atmosphere is very moist and unstable, individual clouds

can grow to great heights, turning into **cumulonimbus** clouds, which we now describe.

6.9.5 Conditional Instability and Cumulonimbus

Consider the temperature profile shown in Figure 6.22, which is a little more complicated than our previous example, and will allow us to introduce an important new concept. The lower portion of the atmosphere is stable; there is even a small temperature inversion that caps the lowest kilometer. Therefore, air parcels lifted from the surface will sink if released and not much will happen, if the lift is small. However, if air parcels are pushed sufficiently high, by a lifting mechanism such as frontal lifting or convergence, and rise above the temperature inversion, they will become positively buoyant and continue to rise. In situations such as this, air parcels are said to be *conditionally unstable* (unstable on the condition that they are lifted past a threshold point – see Box 6.5). **Conditional instability** is an important criterion to consider for thunderstorm development, as it is the most common profile that leads to deep convection; the atmosphere is almost never absolutely unstable over a deep layer.

To illustrate, let us consider an air parcel with temperature $T_p = 16\,°C$ at the surface, and water vapor pressure $e = 5\,hPa$, as in Figure 6.22. Using Figure 5.4, we can determine that its original saturation water vapor pressure is $e_s = 18\,hPa$. We further know that the air parcel will reach saturation when the saturation water vapor pressure decreases to $e_s = e = 5\,hPa$, which will happen at a temperature of $-4\,°C$ (using

Figure 6.21. Cumulus congestus.

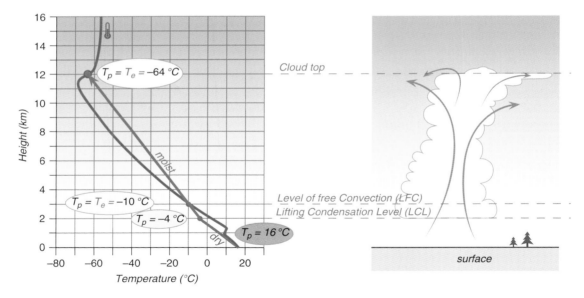

Figure 6.22. (a) Example of a temperature profile (in red) leading to the formation of a cumulonimbus once air parcels have been forced to the level of free convection (LFC). The blue and green arrows represent the decrease in temperature inside an air parcel as it is lifted from the surface: first at the dry adiabatic lapse rate (blue) and then at the moist adiabatic lapse rate above 2 km (green). The air parcel is buoyant up to about 12 km, where it stops rising. (b) Resulting cumulonimbus cloud.

Figure 5.4 again). Therefore, we would need a 20 °C temperature drop, from 16 °C down to –4 °C, to start forming a cloud. Since the temperature first decreases at the dry adiabatic lapse rate (10 °C/km), a 20 °C temperature drop is achieved with a 2 km gain in elevation. Thus, saturation would occur at an altitude of

2 km (the LCL). From then on, further ascent results in the temperature decreasing at the moist adiabatic lapse rate (6 °C/km).

Although a cloud will start forming at 2 km, the air parcel ($T_p = -4$ °C) would still be colder than the environment at that location ($T_e = 10$ °C), and therefore

Figure 6.23. Cumulonimbus cloud flattening against a stable layer, giving it the appearance of an anvil.

heavier than the surrounding air, and thus subject to a downward buoyant force. To form a deep cloud, we need to *force* the air parcels up to that altitude by some forcing mechanism. If such a forcing mechanism is in place, however, be it frontal lifting, convergence, or otherwise, and if an air parcel reaches 3 km, it will become warmer than the environment (see the green arrow in Figure 6.22(a)), become freely buoyant and will continue to rise by itself. Forced convection will give way to *free* convection. This critical altitude is called the **level of free convection** (LFC) and constitutes an important diagnostic of the state of the atmosphere when there is a possibility of thunderstorms. Indeed, if we follow the ascent of our air parcel, we can see that it will keep rising until it becomes once again colder than the environment, which happens at 12 km, near the tropopause. Our cumulus cloud has grown and developed into a full-fledged **cumulonimbus** cloud,

with the likely presence of thunder and lightning – a **thunderstorm**.

Note that our calculations are greatly simplified for the sake of demonstration, and recall in particular that the moist adiabatic lapse rate is not a constant – our value of 6 °C/km is only an approximation to illustrate the basic process. Therefore, not all thunderstorms grow to be 12 km tall. Nevertheless, fully developed cumulonimbus clouds often span the entire troposphere and abruptly stop at the tropopause, where the environmental temperature stops decreasing, giving way to the very stable stratosphere. That is why these clouds often seem to flatten out at the top, giving them the appearance of an anvil (Figure 6.23). Just as it takes time to stop a speeding car, due to its high speed and inertia, some air parcels overshoot the tropopause and penetrate into the stratosphere, which explains why the cloud top is not always completely flat, and can show some overshooting cauliflower heads ("overshooting tops").

To conclude, we have shown that we can form clouds by lifting air parcels, essentially cooling the air adiabatically until it reaches saturation. Stability of the environmental temperature profile affects all lifting mechanisms, but especially the convective mechanism, which depends on the buoyancy of air parcels to "lift themselves." The stability properties of the temperature profile play a key role in determining the types of clouds that form, which are summarized in a "cloud family album" in Appendix 6.1.

Summary

Cloud droplets form by **heterogeneous nucleation**: when saturation is reached, water vapor molecules condense onto **condensation nuclei** – small, airborne, solid particles (**aerosols**) such as dust, pollen, pollutants, and salt particles from the ocean.

Saturation can be achieved by evaporating water and increasing the water vapor content of the air, as in the case of **evaporation fog**, or by decreasing the temperature of the air, and therefore the saturation water vapor pressure, as in the case of **radiation fog**, and the formation of cloud droplets by **adiabatic expansion** and cooling.

The temperature of a rising air parcel decreases because the parcel expands to adjust to the decreasing environmental pressure. Because this expansion is **adiabatic** (i.e., there is no exchange of heat between the parcel and the environment), the energy needed to accomplish this expansion comes from the kinetic energy of air molecules in the parcel, which results in a decrease in temperature of the air parcel. This **adiabatic cooling** explains the formation of many clouds in the atmosphere (except for fog, as well as some cirrus and stratocumulus clouds).

When air is not saturated, a rising air parcel cools at the **dry adiabatic lapse rate**, which is about $10\,°C/km$. When the air parcel is saturated, water vapor condenses to form cloud droplets as the air parcel ascends. Condensation releases latent heat, which warms the air. Therefore, a saturated rising air parcel cools at the **moist adiabatic lapse rate**, which is about $6\,°C/km$.

The altitude at which saturation is reached is called the **lifting condensation level** (LCL). A rising air parcel cools at the dry adiabatic lapse rate *below* the LCL, and at the moist adiabatic lapse rate *above* the LCL.

When air is forced over a mountain (**orographic lifting**), it tends to produce clouds and precipitation on the windward side, and clear weather on the lee side. Because the air cools at the moist adiabatic lapse rate on the windward side above the LCL, and at the dry adiabatic lapse rate on the lee side, the subsiding air on the lee side is both drier and warmer than on the windward side. These lee winds are found in many locations on Earth and given various names by the local populations (e.g., **chinook** on the eastern slopes of the Rocky Mountains in the United States, and **foehn** in the Alps in Europe).

Upward motion leading to the formation of clouds can be achieved by orographic lifting, by **convergence**, by **frontal lifting**, and by **convection**.

An air parcel is said to be **stable** when it returns to its original position after being displaced. It is said to be **unstable** when it accelerates away from the location from where it was displaced, which means positively buoyant and rising when lifted, and negatively buoyant and sinking when lowered. When an air parcel becomes positively buoyant only when it is lifted past a certain threshold, it is said to be **conditionally unstable**.

Temperature inversions are always stable: they inhibit vertical motion, convection, and the formation of clouds.

Lifting of unstable air tends to produce **cumuliform** clouds, whereas lifting of stable air tends to produce **stratiform** clouds.

A **cumulonimbus** cloud typically forms when rising air parcels reach the **level of free convection** (LFC): the level at which the air parcels continue rising under the sole effect of buoyancy, without any other lifting mechanism. Cumulonimbus clouds often span the entire depth of the troposphere. Since the stratosphere is stable (by virtue of the stratospheric temperature inversion), the rising air parcels in the cloud cannot penetrate into the stratosphere, and the cloud flattens out at the tropopause, giving the top part of the cumulonimbus cloud the appearance of an anvil.

Appendix 6.1   A Cloud Family Album

Figure 6.24. Cumulus clouds.

Figure 6.25. Cumulus congestus.

Figure 6.26. Cumulonimbus cloud.

We now know enough about cloud formation to understand the logic behind cloud classification. First, the cloud formation process tells us about the *type* of cloud, and it is convenient to distinguish between cumuliform and stratiform clouds.

Cumulus clouds (Figure 6.24) form in convective situations, and relatively rapid vertical motion with entrainment and mixing often gives them a round, plump, cauliflower appearance. Fair-weather cumulus clouds form when the atmosphere is only mildly unstable. They have a limited vertical extent and a flat base. They are separated by areas of subsidence, and do not produce any precipitation. They are typical of warm, sunny afternoons, when the upper atmosphere is stable.

When the atmosphere becomes more unstable and convection intensifies, or when a lifting mechanism forces the air upward (convergence, for example, or a cold front), the clouds accumulate and grow into **cumulus congestus** (Figure 6.25).

These often block the horizon and are noticeable because of their bright white appearance.

When the atmosphere is very unstable and an individual cloud grows vertically, sometimes spanning the entire troposphere and flattening out against the tropopause, we have a **cumulonimbus**, and most likely thunder and lightning. (Technically, "nimbus" means "rain-producing.") In Figure 6.26, taken by a NASA astronaut on the International Space Station, we see a cumulonimbus from above, with an extensive flat top and some overshooting tops where convection is most intense. The thunderstorm is surrounded by a number of cumulus clouds, including cumulus congestus. Some of the rising cloud towers might themselves grow into more cumulonimbus clouds later during the day.

Stratus clouds have more horizontal than vertical development. (Stratus, in Latin, means

Figure 6.27.   Stratus clouds.

Figure 6.28.   Cirrus clouds.

Figure 6.29.   Altocumulus clouds.

"spread out.") They form when the air rises more gradually – typically, when warm air overruns a cold air mass along a warm front (see Figure 6.12(d)). These clouds can stretch for hundreds of kilometers and effectively block the sunlight. Therefore, their underside is often dark gray (Figure 6.27). When viewed from above, from a mountain or a plane, they appear as a bright white blanket covering the entire landscape. Rain-producing stratus clouds are called **nimbostratus**.

Cirrus clouds form high in the troposphere, where the temperature is less than −40 °C, and are made of ice crystals. These lend a wispy appearance to the cirrus clouds, with no well-defined boundary (Figure 6.28). Through mechanisms we will discuss in Chapter 7, the ice crystals grow and sometimes fall, forming **virga**. When these falling ice crystals are blown by upper-level winds, they give the cirrus clouds the appearance of a veil, of falling hair, or

of an animal's tail. Cirrus, in Latin, means "curl of hair," and such drifting ice particles are sometimes called "mare's tails" or "fallstreaks."

Even though the processes leading to each type of cloud are quite different, the distinction can sometimes sound academic, as several processes can be at work at the same time, and some clouds seem to fit different descriptive types. That is why we introduce another parameter in classifying clouds: the altitude at which they form. We typically distinguish low clouds, midlevel clouds, and high clouds. Midlevel clouds are given the prefix "alto-." Since high clouds are almost always made of ice crystals, they are given the prefix "cirro-."

Thus we can expand our cloud family and distinguish **cumulus** (a low cloud), **altocumulus** (Figure 6.29, a cloud forming in a mildly unstable layer of the midlevel atmosphere), and **cirrocumulus** (Figure 6.30), made of ice crystals. The cloud bundles in altocumulus are closer together than fair-weather cumulus clouds, and bigger in size than in a cirrocumulus – although it is sometimes difficult to distinguish them. Altocumulus clouds are also darker than cirrocumulus, and sometimes share the sky with stratocumulus, while cirrocumulus often coexist with other cirrus clouds, such as cirrus and cirrostratus.

Similarly, we can expand the stratus family and distinguish low clouds (stratus and nimbostratus), midlevel clouds (altostratus), and high clouds

Figure 6.30. Cirrocumulus clouds.

Figure 6.31. Cirrostratus, with halo around the sun.

(cirrostratus). They all have wide horizontal extent, but differ in several respects. **Stratus** clouds are low in the atmosphere, are dark gray and gloomy, and let little sunshine through. **Altostratus** are not as thick, and the sun is usually visible as a faint white disk – although the light is not sufficient to cast strong shadows. **Cirrostratus** are clearly distinct: they have a whitish appearance, the fuzziness typical of cirrus clouds, and definitely appear higher in the sky. The sun is visible and often surrounded by a **halo**, as in Figure 6.31: indeed, as sun rays go through the cloud, they are refracted by ice crystals at a 22° angle specific to the structure of the water crystal (see Chapter 7).

The three types of stratus can sometimes be observed when a warm front is approaching, or, in other words, when a midlatitude cyclone is approaching. If you are standing in the path of the storm, as shown in Figure 6.32, with the warm front advancing toward you, you will see a series of clouds such as shown in Figure 6.33, with low stratus on the horizon, turning into higher altostratus, and finally cirrostratus overhead, possibly mixed with

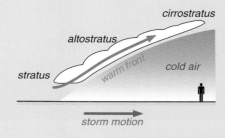

Figure 6.32. Typical cloud sequence associated with a warm front.

cirrus and cirrocumulus. As the front advances, you will find yourself under ever lower and thicker stratiform clouds, and eventually precipitation, as the front moves through. A thickening cirrostratus cloud is a good sign of an approaching storm, and a sign that sailors and farmers knew to recognize when weather forecasting was still empirical.

Stratus clouds sometimes break up into cumulus slabs, and are then called **stratocumulus** (Figure 6.34). They can also be the result of mild convection being impaired by strong subsidence aloft, as could happen under an anticyclone. This is the case for large sections of ocean in the subtropics.

Figure 6.33. Approaching strom.

Figure 6.35. Lenticular cloud.

Figure 6.34. Stratocumulus clouds.

In the end, know that clouds can fall between categories, and it is more important to understand the mechanism behind the formation of the cloud than it is to get its name right.

Finally, remember that any lifting mechanism can lead to cloud formation, given the right circumstances. When wind is deflected above a mountain, for example, a cloud can form if the air is sufficiently close to saturation. When the atmosphere is stable enough that the flow remains horizontal (with no vertical cloud growth), a **lenticular** cloud can form (Figure 6.35).

CHAPTER 7

Precipitation

Why does water fall from clouds? How do cloud droplets turn into raindrops? Why are there different types of precipitation? These are some of the questions we will answer in this chapter, as we explore how cloud droplets, and in particular ice crystals, grow through various processes and turn into precipitating particles. From rain to hail and snow, we will describe specific conditions and processes in which various types of precipitation occur.

Why do some clouds produce precipitation, while others don't? There is a very large size difference between cloud water droplets (on the order of $10\,\mu m$ in diameter) and raindrops (on the order of $1\,mm$ in diameter). To use an analogy, if a cloud droplet were the size of a very small green pea, then a raindrop would be about the size of a basketball. Thus, a cloud droplet needs to grow a hundredfold in diameter, and therefore a *millionfold* in volume, to become a raindrop. While cloud droplets can grow by condensation of additional water vapor at first, this is a relatively slow process, and condensation alone is not sufficient to explain the rapid growth of raindrops in a precipitating cloud. Indeed, additional processes are at work, and the details of the processes depend on whether or not ice particles are present in the cloud.

7.1 Warm vs. Cold Clouds

To describe the full spectrum of precipitation processes, it is useful to define two types of clouds: **warm clouds** and **cold clouds**. In warm clouds, the temperature everywhere is above freezing ($T > 0\,°C$), whereas in cold clouds, there exist parts of the cloud that are below freezing.

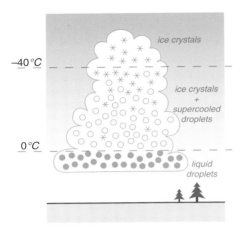

Figure 7.1. Distribution of liquid water droplets (blue), supercooled droplets (white), and ice crystals (stars) in a cold cloud, as a function of temperature.

Warm clouds only contain liquid water droplets and water vapor, whereas cold clouds also contain ice particles. In fact, cold clouds can contain a little of everything depending on altitude and temperature (Figure 7.1). Where the temperature is above freezing, only liquid water droplets and water vapor are to be found (except maybe for falling frozen particles that are melting). Where the temperature is below −40 °C (typically the top of the cloud, or a very high cloud), only ice crystals and water vapor are to be found. A high cloud containing exclusively ice particles, such as a cirrus, is said to be **glaciated**.

What should we expect in between? You might think that only frozen water should be found when the temperature is below 0 °C, but that is not necessarily the case. Liquid water droplets can exist at temperatures below 0 °C. They are said to be **supercooled**, and are well known to aircraft pilots, because they freeze on contact with the surface of the aircraft, or within the engine, and can cause the wing flaps to freeze or the engine to stall. If you have ever been on a plane that was "de-iced" prior to takeoff, it was done so to prevent this from happening. In fact, supercooled liquid water droplets are what we mostly find just above the freezing level, where the temperature is less than 0 °C but still greater than −10 °C.

Then there exists a region of the cloud where supercooled liquid water droplets and ice particles coexist, where the temperature is below −10 °C, but still above

−40 °C. As shown in Figure 7.1, the distribution shifts from a higher concentration of supercooled liquid water droplets to a higher concentration of ice particles as we rise in the cloud and the temperature decreases. A cloud containing both ice particles and supercooled liquid water droplets is called a **mixed-phase cloud**.

Fair-weather cumulus clouds are relatively low in the atmosphere, and in summer are a good example of warm clouds. In contrast, cirrus clouds are high in the troposphere, and are a good example of cold clouds. They only contain ice particles, which gives them their wispy appearance. Cumulonimbus clouds are tall and span the entire range of temperatures shown in Figure 7.1. Their lower portion contains mostly liquid water droplets, which gives them their cauliflower-like appearance, with sharply defined boundaries, whereas their top portion only contains ice particles, which explains why the top of the *anvil* is often more wispy and not as clearly defined.

7.2 Collision and Coalescence

In warm clouds, droplets grow by colliding with each other and coalescing to form larger drops. They do so not only due to the presence of turbulent motions, which make droplets collide randomly, but also because the droplets fall at different speeds. Bigger drops fall faster than smaller drops, because friction has less of a dragging effect on bigger drops, relative to their size and weight. The drag of air molecules against the side of the droplets, as they fall through the air, is larger for smaller droplets (relative to their size), which creates greater resistance and slows the fall speed.

If all cloud droplets had the same size and fell at the same speed, they would never meet, i.e., the **collision efficiency** would be low. If a large droplet were falling through much smaller droplets, the large droplet would fall faster than the very small ones, but the small droplets would be swept aside by the displacement of air caused by the large falling droplet, i.e., the collision efficiency would again be low. If the differences in size are neither too large nor too small, however, the droplets are more likely to collide, i.e., the

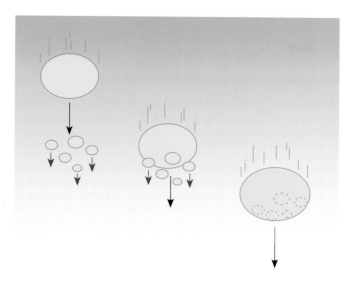

Figure 7.2. Growth of cloud droplets by collision-coalescence as a bigger droplet collects smaller droplets on its way down.

collision efficiency is higher. Because there is typically a wide range of drop sizes in a cloud, they have many opportunities to collide and **coalesce**, i.e., to merge into bigger droplets (Figure 7.2).

Ultimately, the drops become so big that their weight overcomes any updraft and they fall out of the cloud. They can be as small as 1 mm in diameter (drizzle) or as big as 5 mm in diameter (in a thunderstorm, for example).

7.3 Ice-Crystal Growth

In cold clouds, ice particles can form in different ways. They can be the result of the freezing of supercooled liquid water droplets. The freezing can occur by *homogeneous nucleation*: in much the same way as water vapor molecules would need to come together and survive as a stable clump of liquid water to form a cloud droplet, liquid water molecules inside a supercooled droplet need to come together and survive as a clump of ice, an embryo from which the rest of the droplet can freeze. In practice, because the kinetic energy of the water molecules must be low enough for this to happen, homogeneous nucleation occurs preferentially at colder temperatures. For temperatures of

around −35 to −40°C and colder, it happens reliably; there is very little liquid water at these temperatures.

Most often for temperatures between 0 and −35 °C, freezing is seeded by the presence of a **freezing nucleus** inside the supercooled droplet, in somewhat the same way that cloud droplets are seeded by condensation nuclei (heterogeneous nucleation). The freezing nucleus, also called more generally an **ice nucleus**, provides a surface to which the liquid water molecules can attach, making the crystal grow and causing the entire droplet to freeze. The reason liquid water droplets persist at temperatures below freezing is because, unlike condensation nuclei, which are abundant, ice nuclei are relatively rare; they require special properties that promote the formation of the crystalline lattice specific to frozen water.

Cloud droplets can also freeze rapidly by *contact* with an ice nucleus (in the same way as you can cause a tub of supercooled liquid water to freeze rapidly by throwing in a piece of ice).

Once we have an ice crystal, there are several ways that it can grow. One mechanism for growth is *deposition* of water vapor molecules onto an ice crystal. This mechanism was discovered at the beginning of the twentieth century by Alfred Wegener (who also formulated the continental drift hypothesis), Tor Bergeron, and Walter Findeisen. It is usually known as the **Bergeron process**. The three scientists made use of the fact that the saturation water vapor pressure over ice (e_{si}) is *lower* than that over water (e_{sw}). In other words, less water vapor can exist in the presence of ice than can exist in the presence of liquid water, at the same temperature (Figure 7.3). Some values of e_{si} and e_{sw} are shown in Figure 7.4.

To grasp fully the significance of this result, let us imagine the cold part of a cloud where supercooled liquid water droplets coexist with ice crystals, and let us further zoom in onto a liquid water droplet in the presence of an ice crystal (Figure 7.5). The temperature is −15 °C and the saturation water vapor pressure is 1.9 hPa over water and 1.6 hPa over ice. The water vapor surrounding the droplet and the ice crystal cannot be in equilibrium with both at the same time. Let's imagine that it is in equilibrium with the water droplet – that means the water vapor pressure is 1.9 hPa. Since the vapor pressure is greater than 1.6 hPa, the air is supersaturated from the point of view of the

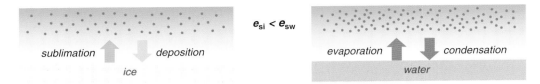

Figure 7.3. Schematic representation of the saturation water vapor pressure over ice and water at the same temperature. When the air is saturated and in equilibrium with the underlying surface, less water can exist as vapor over ice (e_{si}) than over water (e_{sw}).

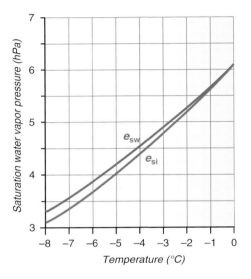

Figure 7.4. Saturation water vapor pressure over water (e_{sw}) and ice (e_{si}) in relation to temperature.

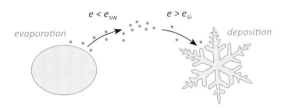

Figure 7.5. Growth of ice crystals at the expense of supercooled liquid water droplets. (Blue dots represent water vapor molecules.)

ice crystal, and, as a result, some water vapor molecules will be deposited onto the crystal, contributing to its growth. But in the process, moisture has been removed from the air, and the water vapor pressure has decreased below 1.9 hPa. Therefore, the air is no longer saturated from the point of view of the water droplet and some water molecules will evaporate from the droplet to saturate the air. Thus, as soon as some water vapor is deposited onto the ice crystal, some water is evaporated from the droplet to compensate the loss. In other words, the ice crystal grows at the expense of the water droplet, which is slowly depleted.

Once initiated, this process continues and allows small ice crystals to grow into large crystals. Additional

mechanisms then contribute to the multiplication and growth of ice crystals. First of all, being relatively fragile, large ice crystals can break into many small parts through **fragmentation**, providing many new ice nuclei that can seed new ice crystals (see later in this section). Second, supercooled liquid water droplets can freeze onto an existing ice crystal during a collision – a process called **riming**, or **accretion**, which tends to form conical spongy pellets called **graupel**. It is thought that ice particles forming by riming might also fragment into numerous ice splinters, providing many new ice nuclei. And third, when two ice crystals collide, they can simply *stick* to each other and form a larger crystal by **aggregation**, which is how large snowflakes are created.

Depending on the temperature, ice crystals of different shapes can form, as summarized in Box 7.1 after results from controlled laboratory experiments. Although very different in appearance, they all share the same six-sided configuration. This is due to the very unique angular structure of the water molecule, which favors a certain orientation and organization of the molecules in the crystals. In actual clouds,

Box 7.1. Ice Crystals

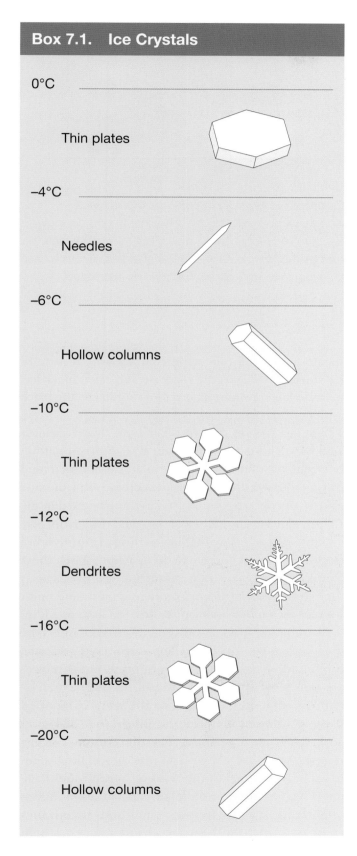

0°C

Thin plates

−4°C

Needles

−6°C

Hollow columns

−10°C

Thin plates

−12°C

Dendrites

−16°C

Thin plates

−20°C

Hollow columns

ice particles appear to be more irregular in shape. Their structure depends on the environmental conditions (temperature, moisture content, strength of updrafts, turbulence, etc.), and can also be determined by the particular ice nucleus that seeded the ice crystal. The hexagonal shape of ice crystals is key to the formation of certain optical phenomena (see Appendix 7.1).

As we mentioned previously, ice nuclei are relatively rare, as they need to have a very specific structure. The best ice nuclei are in fact ice crystals themselves, or bits and pieces created when a crystal fragments. Engineers, however, have tried to manufacture ice nuclei, hoping to seed the atmosphere and to provoke precipitation, or to affect the size distribution of the precipitating particles (one of the first attempts at **weather modification**). In 1946, Vincent Shaefer, a General Electric researcher, discovered that dry ice (frozen carbon dioxide) could be used to modify clouds. The dry ice creates ice crystals which then grow at the expense of the liquid water. A short time later, Schaefer's colleague, Bernard Vonnegut, discovered that silver iodide had a structure similar to that of ice and should also modify the distribution of ice crystals in clouds. Since then, many experiments have been carried out, but available statistical analyses are inconclusive and the contribution of silver iodide to the formation of clouds and precipitation remains unclear.

7.4 Precipitation Types

Why do we experience different types of precipitation? There are a number of factors that can affect the type and size distribution of precipitating particles. Temperature, of course, is one such factor, but the dynamics of the weather system in which precipitation forms play a more important role in the end. In warm clouds of the convective type (cumulus, cumulus congestus, cumulonimbus), droplets grow by collision and coalescence and fall as "fat" raindrops in the form of **showers**, possibly quite intense, and rarely long lasting, unless there is a forcing mechanism that maintains the convective cloud in place. When convection is very intense, especially if the thunderstorm

Figure 7.6. Hailstones can grow to a variety of sizes as they go through different regions of the cloud and accumulate ice.

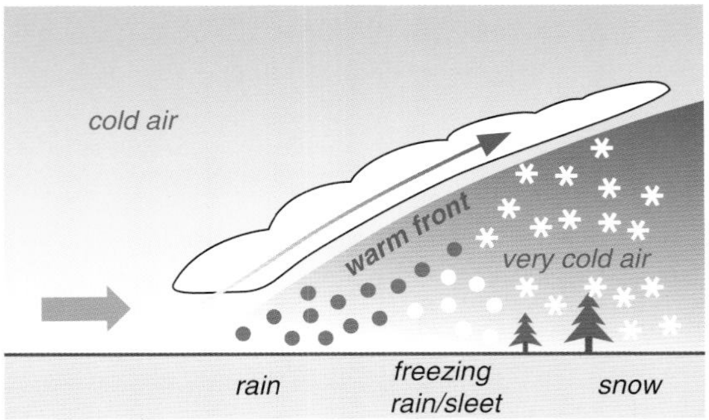

Figure 7.7. Snow falling off a stratiform cloud along a warm front. Ice crystals might persist as snow deep into the very cold air mass, but they might also melt as they traverse the warmer air mass, in which case they can form rain, or refreeze to form either sleet or freezing rain.

rotates, separating updrafts from downdrafts (see Section 11.4), droplets can be pushed up high into very cold regions of the cloud, where they freeze. As they leave the updraft and fall through the cloud, they accumulate ice by riming and grow in size, and then cycle back up into the updraft to freeze before adding another layer. These little balls of ice eventually become too heavy for the updraft to support them and they fall to the surface, where we experience them as **hail** (Figure 7.6).

When the updraft is relatively gentle and in shallow clouds, the size range of droplets is small. As a result, droplet collisions are not as frequent and raindrops remain relatively small. When raindrops are very small, we refer to them as **drizzle drops**.

In the midlatitudes, and especially in winter, the upper atmosphere is usually much below 0 °C and precipitation often starts as falling ice crystals. If they remain frozen all the way to the surface, we experience **snow**. If they melt, however, as they fall through a deep layer of warmer air, we experience rain. Sometimes, they melt as they fall through a first layer of warmer air, but refreeze as they continue down through a deep colder air mass near the surface. In this case, we observe small pellets of ice, which we call **sleet** (Figure 7.7). When the ice crystals melt as they fall through warm air, and fall through a shallow cold layer only to refreeze when they touch the ground, we

speak of **freezing rain** – the characteristic feature of **ice storms**. When on the road, we also call it **black ice**, because it is transparent and looks similar to the black asphalt underneath.

The last type of rain you may have heard about is **acid rain**. This is not a different type of precipitation, but regular raindrops containing sulfur-based and nitrogen-based dissolved gases such as sulfur dioxide (SO_2), nitrogen oxide (NO), and nitrogen dioxide (NO_2). These gases, blown by the wind from the exhaust of power plants located upstream, dissolve in cloud droplets and make them acidic. When these cloud droplets grow into raindrops and precipitate, they contaminate the environment, plants, and water sources (see more on air pollution in Chapter 14).

As for snow, there are three basic ingredients conducive to snow formation: moisture, cold temperatures, and of course upward motion. If temperatures are quite low, the snowflakes will be relatively small and dry and we then refer to the snow as powder. If the conditions are also very windy, then the snow contributes to a **blizzard**. On the contrary, if the temperatures are above 0 °C, the snowflakes begin to melt, which makes them more "sticky" and facilitates aggregation. Snowflakes can fall about 300 m (1000 feet) in above-freezing conditions before they completely melt, with surface temperatures as high as about 5 °C.

Summary

As far as precipitation is concerned, it is convenient to divide clouds into two types, warm and cold.

▶ **Warm clouds**, in which the temperature is everywhere above freezing ($T > 0\,°C$), contain exclusively liquid water droplets and water vapor.

▶ **Cold clouds** contain parts that are below freezing. Cold clouds can contain liquid water droplets, **supercooled droplets** (below $0\,°C$, but not frozen), **ice crystals**, and water vapor.

A high cloud containing exclusively ice particles, such as a cirrus, is said to be **glaciated**.

To form raindrops, cloud droplets must grow from about one micrometer to at least one millimeter in size. In a warm cloud, they grow by **collision** and **coalescence** of cloud droplets falling at different speeds. In a cold cloud, several processes are at work:

▶ Supercooled liquid water droplets in high clouds, with temperatures below −40 °C, can freeze by homogeneous nucleation (i.e., from water molecules alone).

▶ Freezing can be initiated by the presence of a **freezing** (or **ice**) **nucleus**.

▶ Ice particles can then grow by the **Bergeron process**, in which ice crystals grow at the expense of liquid water droplets by deposition of vapor onto the ice crystal due to a difference between the respective saturation water vapor pressures over ice and water.

▶ Large ice crystals can break into many small parts through fragmentation, providing many new ice nuclei that can seed new ice crystals.

▶ Supercooled liquid water droplets can freeze onto an existing ice crystal during a collision (**riming**, or **accretion**), forming conical ice pellets called **graupel**.

▶ Two ice crystals can collide and stick to each other to form a larger crystal by **aggregation**.

Precipitation can take the form of:

▶ **Rain showers**, when droplets grow by collision and coalescence, typically in a convective cloud, and fall in a relatively short amount of time. They may also originate as snowflakes and melt before reaching the ground, and continue to grow by collision and coalescence on their fall through the warm part of the cloud.

▶ **Hail**, when droplets are pushed up high into very cold regions of a convective cloud, freeze, and sweep repeatedly in and out of the updraft of the cloud, accumulating successive layers of ice by riming.

▶ **Light rain**, when the updraft is expansive and relatively gentle, as happens along a warm front.

▶ **Drizzle**, when raindrops are very small.

▶ **Snow**, when falling ice aggregates remain frozen down to the surface. Snow can have a wide range of textures, from small and powdery to large, wet snowflakes.

▶ **Sleet**, or ice pellets, when rain freezes in a deep layer of cold air near the ground.

▶ **Freezing rain**, when rain falls through a shallow layer of cold air at the ground, freezing in contact with features near the surface (the ground, trees, etc.).

Appendix 7.1   Some Optical Phenomena

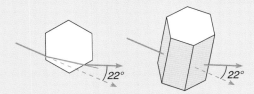

Figure 7.8. Double refraction of light passing through an ice crystal.

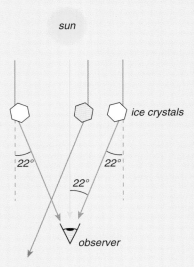

Figure 7.9. Sun rays refracted by ice crystals, as seen by an observer.

The interaction of sunlight with cloud and precipitating particles can give rise to various optical phenomena, two of which we describe here. We encountered the first one in Chapter 6: the **22° halo** observed around the sun in the presence of a cirrostratus cloud (see Figure 6.31 in Appendix 6.1). Such a halo is caused by the double refraction of sunlight as it passes through hexagonal ice crystals (Figure 7.8). An observer facing the sun will see, in addition to direct sunlight, those rays that are refracted by ice crystals in such a way as they reach the observer's eyes after a 22° refraction (see the red lines in Figure 7.9), but will not see the sun rays refracted by the other ice crystals (gray line in Figure 7.9). Therefore, when refraction takes place all around the sun, the observer sees a ring of light at a 22° angle from the sun, i.e., a circular halo (Figure 7.10). Halos can also be observed around the moon, as shown in Figure 7.11. When the sun's rays are refracted through the top or base of the ice crystals (rather than the sides), they can cause a 46° halo, although it is more rare and difficult to observe.

When light passes through *liquid* raindrops, a **rainbow** can form. In this case the sun's rays are first refracted as they enter a droplet (*a* in Figure 7.12(a)), then are reflected by the back of the droplet (*b*), before finally being refracted again on the way out (*c*). The rays exit the droplet in the reverse direction that they entered, at a 40–42° angle from the direction of the original sun ray.

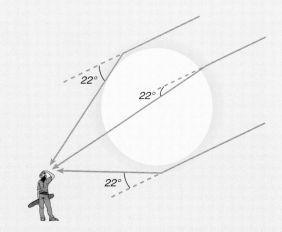

Figure 7.10. Circular halo resulting from the refraction of sunlight passing through ice crystals.

Moreover, sun rays of shorter wavelength are refracted at a greater angle than those of longer wavelength, which results in a separation of colors as they exit the raindrop (Figure 7.12(b)). Therefore, an observer standing with their back to the sun watching raindrops falling will see different colors

Figure 7.11. Halo around the moon.

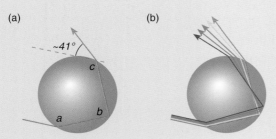

Figure 7.12. Refraction and reflection of light passing through a water droplet. (a) Angle combination. (b) Wavelength-dependent angles.

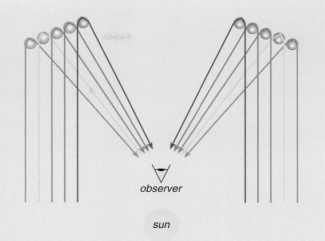

Figure 7.13. Different colors emerging out of different raindrops, as seen by an observer.

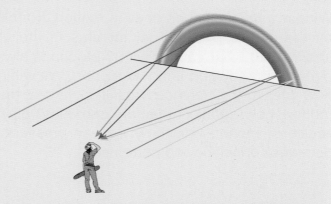

Figure 7.14. Rainbow resulting from wavelength-dependent refraction of sunlight passing through raindrops.

Figure 7.15. Rainbow.

emerging out of different raindrops, as shown in Figure 7.13.

Of course, each raindrop also refracts and reflects other colors, but these leave the raindrop at a different angle and the observer cannot see them. The band of raindrops that refract and reflect sun rays toward the observer is relatively narrow and semi-circular, and takes the form of a rainbow (Figure 7.14 and 7.15).

CHAPTER 8

Wind

Wind is a manifestation of air in motion, from light breezes by the sea to powerful jet streams in the upper troposphere. But what causes wind? Motion is always initiated by forces, and wind is no exception. We will now explore how pressure differences result in forces that set air in motion, how additional factors come into play at different scales of motion, and how a given pressure distribution generates specific wind patterns.

CONTENTS

In trying to understand the formation of wind, we must return to the fundamental laws of physics, and in particular the law of conservation of momentum, in the form of Newton's laws of motion. These will allow us to have both a **diagnostic** approach (i.e., describing the current state of the atmosphere) and a **prognostic** approach (i.e., predicting the future state of the atmosphere – see Chapter 13).

8.1 Force and Acceleration

As for any moving object, air follows the laws of motion described by Isaac Newton (1643–1727) in the seventeenth century. Newton's first law states that

An object at rest remains at rest, and an object in motion moves in a straight line at constant speed unless acted upon by a force.

Newton's second law states that

> The acceleration experienced by an object is equal to the force exerted on that object divided by its mass.

This is captured by the famous expression $\mathbf{F} = m\mathbf{a}$, or $\mathbf{a} = \mathbf{F}/m$, where an object of mass m experiences an acceleration $\mathbf{a}$ when a force $\mathbf{F}$ is applied to it. [Bold letters indicate that the quantities are **vectors**, with both direction and magnitude.] Here the air has a certain velocity (i.e., speed and direction), and the acceleration is the rate of change of that velocity, in the same way as the speed of a car changes when we push on the accelerator. Since the acceleration is equivalent to the force applied to the air divided by its mass, understanding wind amounts to understanding the forces applied to the air in various circumstances. When there is more than one force, Newton's laws apply to the *net* force, which is the sum of all forces.

In describing air motion, it is useful to differentiate vertical from horizontal motion. Vertical motion is very much constrained by gravity and is described by buoyancy and stability, as explained in Chapter 6. Horizontal motion is free of gravity and is determined by a set of horizontal forces: the pressure gradient force, the Coriolis force, the centrifugal force, and friction, which we will now describe.

8.2 Pressure Gradient Force

Consider the following thought experiment, in which a glass of water has been flipped over on a table top (Figure 8.1(a)). As we lift the glass, water flows out and spreads over the table top. Why does the water flow outward? We often hear that water "wants to equalize" and be level, but what is the force that sets the water in motion to make that happen? For the glass of water, we know that, right at the counter top, where the water column is highest, the pressure is also higher than outside the glass because the water column weighs more than a column of air outside the glass. Therefore, there exists a *pressure difference* between the interior and the exterior of the water column (Figure 8.1(b)). In dealing with wind, we prefer to use the term **pressure gradient**, since it is a vector that has both magnitude and direction.

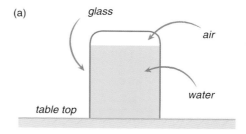

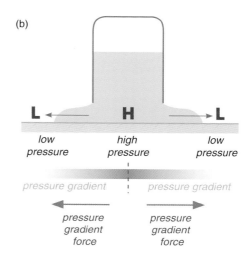

Figure 8.1. Thought experiment, in which water spreads out over a table top as it falls from an inverted glass.

Because there is a pressure gradient, more active water molecules at higher pressure push outward against less active molecules under less pressure. In other words, the pressure difference creates a *force* directed outward from the water column that pushes the water out. We call this force the **pressure gradient force** (PGF), and note that it is *always directed from high toward low pressure.*

If we change our point of view to a bird's eye view, we obtain a good analogy for the mechanism that initiates motion in the atmosphere, as illustrated in Figure 8.2(a). When there is a region of high pressure surrounded by lower pressure, the pressure gradient force pushes the air outward, causing wind to blow. In the same way that the water spreads out and away from the glass, we can picture the air spreading out and away from the high pressure center, and emptying the air column. In fact, the analogy extends to the vertical: in the same way that the water level falls

(a)

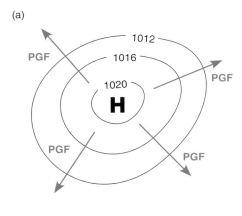

(b)

Figure 8.2. Distribution of pressure gradient force (PGF) around a high pressure region (a) and a low pressure center (b).

in the glass, air tends to subside in a high pressure region, to replace the surface air exiting through the sides.

Although it is more difficult to create such a straight analogy for low pressure centers with a glass of water (we have to create a partial vacuum to create a column of air that weighs less than the surroundings), we can use our imagination and flip the above picture to obtain Figure 8.2(b). The pressure gradient force pushes inward, from high to low pressure, and wind will blow toward the low pressure center. Of course, other forces will play their part to create a slightly more complicated picture, and winds will end up blowing, not in and out of lows and highs, but *around* lows and highs, as we will soon explain. But in the meantime, the pressure gradient force is the first force we should consider when trying to understand the origin of wind, and indeed the main one when trying to understand small wind circulations covering

short distances and time periods. To illustrate, we will take the example of a sea breeze on the coast.

8.3 Sea Breeze and Land Breeze

Imagine a sunny day at the coast where we can observe the coastline. As we explained in Chapter 4, land warms and cools much more and much faster than water (due to the larger heat capacity for water, the redistribution of heat in water by mixing, the vertical spreading of heating in water due to the penetration of light to greater depths, and the evaporation of water requiring latent heat). Uneven heating of the surface carries over to the two air columns over the land and ocean, so that, during the day, the lower part of the land air column is warmer than the lower part of the ocean air column. Now recall that warm air expands, and, in our particular case, the land air column expands upward to greater heights than the ocean air column.

Let us draw some pressure levels and observe how they change due to this uneven expansion (Figure 8.3(a)). Pressure surfaces are raised higher over the land than over the ocean, due to thermal expansion, which makes them tilt as shown in Figure 8.3(b). Now consider some horizontal level and assess the pressure gradient. By interpolating between pressure levels, we can estimate that the atmospheric pressure at point *A* in Figure 8.3(b) is about 910 hPa, whereas it is about 895 hPa at point *B*. Therefore, due to the differential thermal expansion and the tilting of the pressure levels, there exists a pressure difference, or pressure gradient, that will cause a wind to blow *aloft* from the land to the sea.

As air is blown from the land to the sea, the air column over the land is depleted of air, while air accumulates in the air column over the sea. Since atmospheric pressure at the surface is tantamount to the weight of the air column above the surface, this redistribution of mass increases the surface pressure over the sea and decreases it over the land, as illustrated in Figure 8.3(c) by comparing the pressure at points *C* and *D*. This surface pressure gradient causes the wind to accelerate, blowing from the sea to the land. Because this wind is usually quite mild, we call

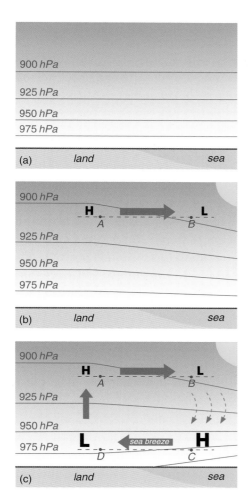

Figure 8.3. Formation of a sea breeze on a sunny day.

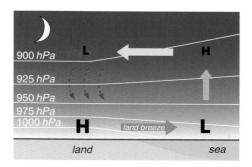

Figure 8.4. Formation of a land breeze at night.

it a *breeze*, and, because it blows *from* the sea, we call it a **sea breeze**.

We note two important points in Figure 8.3. First, the exchange of mass between the two air columns is usually accompanied by rising motion over the land and subsidence over the sea, although subsidence is usually less concentrated and more spread out over the water. This closes the loop and constitutes what we call a **sea breeze circulation**. It is driven by the temperature contrast between land and water, and it occurs in response to the corresponding energy imbalance: it redistributes energy from the warm region to the cold region to bring the system back to a more balanced state.

Note, however, that, even though we indicated a low pressure at ground level over the land and a

high pressure aloft, the surface pressure at D (about 970 hPa) is still much higher than the pressure aloft at A (about 910 hPa). That is why we used a larger **H** and **L** at the surface and smaller ones aloft, to avoid the temptation to diagnose some air flow from A to D or from C to B. Values of atmospheric pressure must be compared on the same horizontal level.

Second, the rising air over the land can lead to the formation of cumulus clouds, while subsidence over the water inhibits cloud formation. Therefore, if clouds form, they will line up along the coast, over the land, while the nearby water will be cloud free – a common situation in summer that can easily be observed from a plane following a coastline.

What happens when the sun sets? There is no longer an input of solar energy into the ground or water, but they both continue to lose energy in the form of infrared radiation, as we explained in Chapter 4. For similar reasons to those given earlier, the land loses energy much faster than the water, and the whole circulation is reversed to give offshore flow, called a **land breeze** (Figure 8.4). If clouds form, they will also line up along the coast, but over the water, and they will be more difficult to observe at night.

This type of circulation can form anywhere there exists weak ambient wind and a temperature contrast – by a lake, for example, or between two types of land surface where one might be a forest, or an expanse of snow. In the end, an important message is that a temperature contrast induces a horizontal pressure contrast, and the resulting pressure gradient causes wind to blow. Moreover, the stronger the temperature contrast, the stronger the pressure gradient, and therefore the stronger the wind.

8.4 Coriolis Force

If the pressure gradient force were the only force acting on the air, winds would blow from high to low pressure. But simple observation of weather maps reveals that this is rarely the case. While it is somewhat true over short distances, such as in the case of a sea breeze, it does not hold over larger distances. Recall the counterclockwise circulation of winds in Figure 2.1. In fact, by comparing Figures 2.1 and 2.4, we can see that the wind does not blow *across* the isobars (perpendicularly) from high to low pressure, as suggested above, but *along* the isobars (parallel), 90° away from what we might expect based on the pressure gradient force alone. It turns out that, after a few hours, Earth's rotation plays a significant role in the wind direction. Considering two points in the direction of the pressure gradient, the wind is accelerated from one point to the other due to the pressure gradient force, but both points are also rotating around Earth's axis. This creates a complicated motion that can only be fully described by taking into account Earth's rotation (i.e., the rotation of the frame of reference from which we observe the motion). The traditional approach to this situation consists in deriving the equations of motion on a rotating sphere, from which we can infer the existence of an acceleration that deflects the air at a right angle to the direction of motion, to the right in the northern hemisphere and to the left in the southern hemisphere. We can think of this effect as either an acceleration (i.e., the Coriolis acceleration) or a force (i.e., the **Coriolis force**, after the French mathematician Gaspard-Gustave Coriolis (1792–1843) who first investigated the matter), depending on our point of view. From that second point of view, the air is accelerated *as if it were pulled* by a force pointing to the right of the direction of motion (in the northern hemisphere). Since that force is not a push or a pull in the traditional sense – it exists only because we view the motion from the rotating Earth – we think of it as an *apparent* force. Mathematically, we

can also determine that the Coriolis force varies with latitude (it is zero at the equator and strongest at the poles) and is proportional to wind speed (although it does not affect the wind speed, only the wind direction).

While such a calculation is beyond the scope of this book, one can get a glimpse of the effect of a rotating frame of reference by considering the following popular analogy. Imagine two persons throwing a ball on a merry-go-round. If the merry-go-round is not spinning, then the ball follows a straight line from the thrower to the catcher (Figure 8.5(a)). If the merry-go-round is spinning (counterclockwise in Figure 8.5), the ball still moves in a straight line after it is thrown, as Newton's first law says it must, which is exactly what we see *if we observe it from outside the merry-go-round* (Figures 8.5(b)–(e)). However, if we observe the ball from the merry-go-round, something strange happens.

If we position ourselves on the merry-go-round, we rotate with it, and it looks to us as if the platform, the thrower, and the catcher are not moving. We are in a rotating frame of reference, and from that point of view we are not moving. If we track the position of the ball from the rotating frame of reference (Figure 8.5(f)), it looks like the ball follows a curved path, as if it were pulled by a force to the right of the direction of motion (for this particular direction of rotation of the merry-go-round, i.e., counterclockwise).

Note the use of the expressions "it looks like" and "as if." It is only an *apparent* deflection of the motion *if you observe that motion from the rotating frame of reference.* The fundamental reason is that the circular rotation of the merry-go-round provides an acceleration to the observer on the platform: if we direct our arm in a certain direction from the merry-go-round, the location we point to changes. Thus, the ball, which is not attached to the rotating merry-go-round any more, appears to be accelerated to the right, but only because we have introduced this apparent acceleration by chosing a rotating point of view.

The deflection of a parcel of air moving on Earth is somewhat similar to the deflection of the ball moving over the merry-go-round. Of course, the analogy does not fully explain the Coriolis acceleration of air and water on Earth. For one thing, Earth is a sphere,

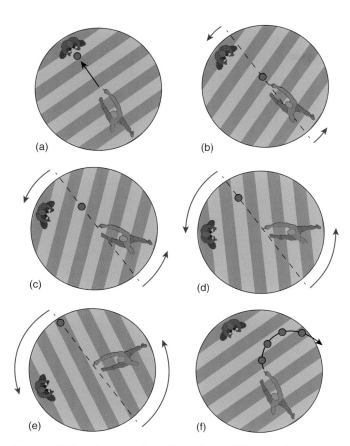

(a)

(b)

(c)

(d)

(e)

(f)

Figure 8.5. Apparent motion of a ball thrown from a merry-go-round spinning counterclockwise. (a) If the merry-go-round is still, the ball (in red) goes in a straight line. (b) If the merry-go-round is spinning, the thrower and catcher rotate with it, but the ball follows the original straight line (dashed) as soon as it leaves the hand of the thrower. (c)–(e) As the merry-go-round continues to spin, the ball is further to the right of the thrower, rather than in front of him. (f) If we position ourselves on the rotating frame of reference and rotate with the thrower and catcher, it appears as if the merry-go-round is not moving and the ball follows a curved trajectory to the right.

not a flat disk. Second, air and water are not free to move out of the rotating frame of reference – if they were, they would fly out to space. They are held down to Earth by gravity, which exerts a very strong constraint. The merry-go-round analogy, however, at least conveys the notion of *apparent motion* and *apparent force* when observing from a rotating frame of reference.

Box 8.1. The Coriolis Force

Rule set #1

(1a) Objects in motion in the northern hemisphere are deflected at a right angle and to the right of the direction of motion.

(1b) Objects in motion in the southern hemisphere are deflected at a right angle and to the *left* of the direction of motion.

(2) The Coriolis force is zero at the equator, and increases with latitude, to a maximum at the poles.

(3) The Coriolis force is proportional to the wind speed: the faster the wind, the stronger the deflection.

(4) Because the Coriolis force is perpendicular to the wind direction, it only affects the wind direction and does not affect the wind speed. In other words, it makes the wind veer, but it does not accelerate or decelerate it.

We always observe the atmosphere from this rotating frame of reference attached to Earth, so we must live with this apparent force. It turns out to play an important role; that is, the fact that Earth is rotating rapidly is fundamentally important to the development of weather. In practice, meteorologists establish basic rules of atmospheric motion that incorporate the Coriolis force, and then apply them systematically.

What are these rules of atmospheric motion? They are very few, in reality, and should be memorized for future reference. We will start with a first set of rules summarized in Box 8.1, and will now examine how the Coriolis force interacts with the pressure gradient force to create specific wind patterns.

8.5 Geostrophic Wind

Let's consider an air parcel embedded in a pressure gradient and observe what part the Coriolis acceleration plays in the development of motion (Figure 8.6).

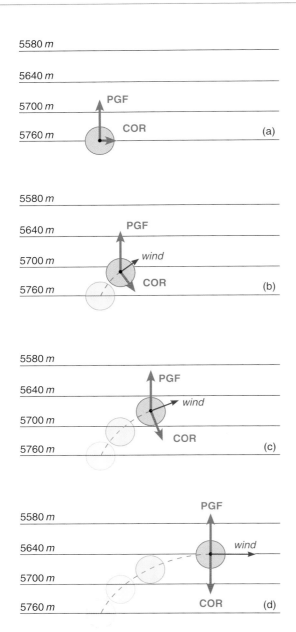

Figure 8.6. Adjustment of an air parcel to geostrophic balance between the pressure gradient force (PGF) and the Coriolis force (COR).

To keep it simple, we choose the midtroposphere and a few straight isobars, parallel to each other and oriented west–east, so that the pressure gradient is oriented from south to north. Recall from Chapter 2 that we can switch interchangeably between pressure contours (at constant height) and height contours (at constant pressure) to observe and understand the motion. In Figure 8.6(a), the height contours decrease

northward from 5760 to 5580 m, indicating that the pressure gradient force (PGF) is pushing air from south to north. For the sake of demonstration, let us imagine that the air parcel is first at rest – a somewhat unrealistic situation, but one that will serve our purpose. Since a force is applied to the parcel (the pressure gradient force), it is accelerated (according to Newton's second law of motion). As soon as it starts to move, it acquires some speed, which means that, from our frame of reference, it is also accelerated to the right of the direction of motion by the Coriolis force (recall that the Coriolis force is proportional to wind speed). Therefore, rather than moving due north, the parcel starts turning to the right (Figure 8.6(b)). Since the direction of motion of the parcel has changed, the Coriolis force changes direction accordingly in Figure 8.6(b) so that it always remains perpendicular to the direction of motion.

As the air parcel continues to move and to accelerate, the Coriolis force increases and keeps deflecting the parcel to the right (Figure 8.6(c)), until the air is blowing due east, at which point the Coriolis force is directed *opposite* the pressure gradient force and the air parcel has acquired enough speed that the Coriolis force is equal in *magnitude* to the pressure gradient force (Figure 8.6(d)). We have reached a state of equilibrium, called **geostrophic balance**, in which there is a balance between two forces equal in magnitude and opposite in direction, such that the air parcel continues to move at a constant speed. Note that the net force applied to the air parcel is now zero – the pressure gradient force and the Coriolis force exactly cancel. Therefore, there is no acceleration, and the parcel keeps moving in a straight line, at the same speed, "unless acted upon by a net force" (according to Newton's first law).

What will happen, however, if the balance of forces changes? If the pressure gradient force increases or decreases? Let us consider the situation represented in Figure 8.7(a), in which the air parcel enters a region where the pressure gradient increases (i.e., the isobars, or height contours, are closer to each other). Since the pressure gradient force increases, the air parcel should be deflected northward (dark blue arrow in Figure 8.7(a)). However, since there is now a net positive force pushing the air parcel

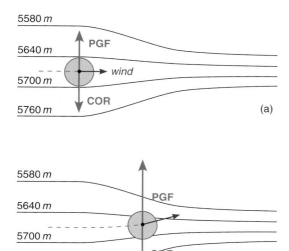

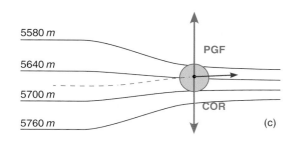

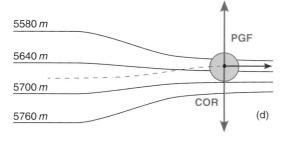

Figure 8.7. Adjustment of geostrophic balance to a stronger pressure gradient.

Box 8.2. Geostrophic Balance

Rule set #2

(1) Wind blows *parallel* to the isobars (i.e., *along* the isobars) with higher pressures on the right (in the northern hemisphere).

(2) Wind speed is proportional to the pressure gradient (i.e., proportional to the tightness of the isobars).

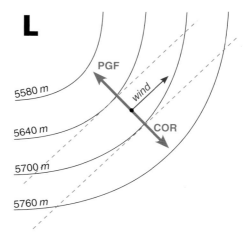

Figure 8.8. In curved flow, we can make a geostrophic balance approximation and assume that the wind is blowing along the tangents (dashed blue lines) to the isobars.

northward, the parcel is also *accelerated* in that direction and its speed increases. The Coriolis force, being proportional to wind speed, also increases and pulls the air parcel southward (Figure 8.7(c)). Eventually a new equilibrium is reached where the pressure gradient force and the Coriolis force are both stronger and balanced, and the wind speed has increased (Figure 8.7(d)).

With this in mind, we can now enunciate two new rules of motion that will help us quickly identify wind patterns using a pressure map (Box 8.2). Given a set of isobars (or, equivalently, height contours), we can use these rules quickly to estimate the direction and relative speed of the wind.

Of course, isobars are not always straight. Most of the time they are curved, as wind blows around highs and lows, troughs and ridges. We can, however, make an approximation and assume that, locally, the wind is blowing in a straight line for a very short time. It amounts to estimating that the wind blows parallel to the local *tangents* to the isobars (Figure 8.8). This will allow us to formulate a general sense for the wind

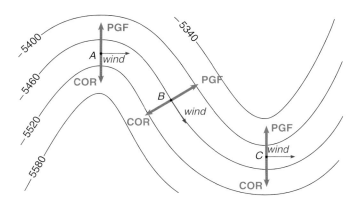

Figure 8.9. Determining the pressure gradient force (PGF), Coriolis force (COR), wind speed, and wind direction from a height contour map assuming geostrophic balance.

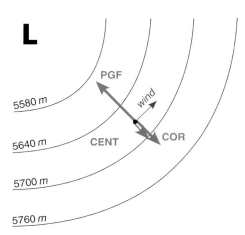

Figure 8.10. Balance of forces around a low when taking into account the centrifugal force (CENT): the wind speed is reduced, the wind is subgeostrophic.

circulation. We should remember, however, that geostrophic balance is only an approximation, and that it works best for straight flow. (See Section 8.6 for a more exact treatment of curved flow.)

Given a pressure map, then, and assuming geostrophic motion, we shall follow the steps summarized in Box 8.3, at whatever point we need to estimate wind speed (in a relative sense) and wind direction. Using this simple methodology, we will obtain approximate wind maps such as shown in Figure 8.9. Note how, in this particular example, high pressures (and heights)

are to the south and low pressures are to the north, and how the pressure gradient force always points from high to low pressure, and is perpendicular to the isobars. The Coriolis force is pointing in the opposite direction to the pressure gradient force, and the wind is blowing along the isobars, first around a ridge of high pressure (point *A*), and then around a trough of low pressure (point *C*).

8.6 Gradient Wind

How good is the geostrophic approximation? It is quite good outside of the tropics when the flow is straight. When the flow is curved, however, the air is subject to an additional "apparent" force (again, like the Coriolis force, this force is due to our choice of reference frame, in a curved flow): the centrifugal force – recall the centrifugal force pulling geostationary satellites away from Earth (Chapter 1) and the centrifugal force you experience when you turn at high speed with your car (apparent because, while your car is turning, your body is trying to move in a straight line). Around a low, the centrifugal force pulls the air in the same direction as the Coriolis force, and it is the sum of both forces that balance the pressure gradient force (Figure 8.10). Therefore, a smaller Coriolis force is required, which corresponds to a smaller wind speed. As a result, the wind speed is less than geostrophic balance would suggest: the wind is said to be **subgeostrophic**. Compare Figure 8.10 with Figure 8.8 to convince yourself that this is indeed the case.

In contrast, around a high, the centrifugal force points in the same direction as the pressure gradient force (Figure 8.11). Therefore, a stronger Coriolis force is required to balance the sum of both pressure gradient and centrifugal forces. As a result, the wind speed is greater than geostrophic balance would suggest: the wind is said to be **supergeostrophic**.

In practice, this new subtlety gives us slightly different wind speed patterns as compared to the geostrophic wind, but the wind direction remains the same: following pressure (or height) contours, with high values on the right (with the wind at your back) in the northern hemisphere. Figure 8.12 shows the same pressure

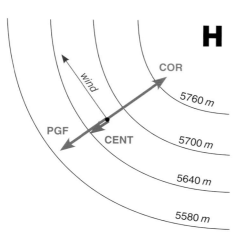

Figure 8.11. Balance of forces around a high when taking into account the centrifugal force: the wind speed is increased; the wind is supergeostrophic.

Rule set #3

(1) Identify the direction of the pressure gradient force.

(2) Knowing that the Coriolis force is equal in magnitude and opposite in direction to the pressure gradient force, draw the Coriolis force.

(3) Knowing that the Coriolis force is pointing to the right of the direction of motion (in the northern hemisphere), identify the wind direction.

(4) Wherever the spacing of the isobars or height contours is wide, reduce the wind speed; wherever it is narrow (i.e., isobars close to each other), increase the wind speed.

pattern as in Figure 8.9, but with different wind speed and Coriolis force once curvature is taken into account. By comparison with Figure 8.9, note how the wind is supergeostrophic around the ridge (point *A*), whereas it is subgeostrophic around the trough (point *C*). Where the flow is relatively straight (*B*), the balance of forces is unchanged from the geostrophic case. Thus,

Rule #4

The wind is supergeostrophic around a ridge and subgeostrophic around a trough.

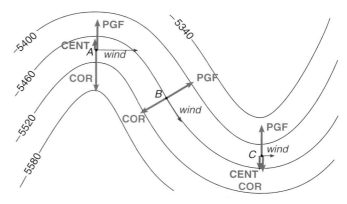

Figure 8.12. Determining wind speed and direction from a height contour map assuming gradient wind balance.

for curved flow, we add a new rule of thumb, summarized in Box 8.4, that will prove particularly useful when analyzing upper-level flow and jet stream patterns.

Note that, in the example shown in Figure 8.12, we took great care to keep all pressure gradients the same as in Figure 8.9, for illustration purposes. In reality, of course, the pressure gradient changes constantly as does flow curvature, yielding non-trivial results (see, for example, Figure 2.10). Furthermore, variations in wind speed create regions of convergence and divergence. These regions are crucial for understanding the development of rising and sinking air, and midlatitude cyclones, as we will see in Chapter 10.

8.7 Surface Winds

The gradient wind approximation is a relatively good approximation aloft (850 hPa and above), and it is tempting to apply it closer to the surface, when

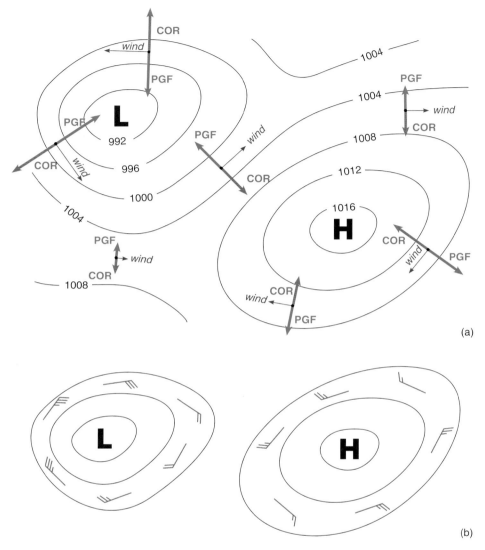

Figure 8.13. (a) Balance of forces and wind pattern at the surface assuming geostrophic balance. (b) Cyclonic circulation around a low vs. anticyclonic circulation around a high (northern hemisphere).

analyzing weather maps at sea level. As we approach the surface, however, yet another factor comes into play: surface friction, which has a more significant impact on the wind pattern than the centrifugal force. So we will put the gradient wind aside for now, and start our discussion of surface winds by returning to the geostrophic approximation.

Figure 8.13(a) shows an example of a surface pressure map featuring a high pressure center (a "high") and a low pressure center (a "low"). Even though the geostrophic approximation is often inadequate close to the surface, applying our rules of geostrophic

motion at several points around the high and the low allows us to see a general pattern emerge. Because winds blow parallel to the isobars with higher pressures on the right, they tend to blow clockwise around high pressure regions and counterclockwise around low pressure centers (in the northern hemisphere) – summarized in Box 8.5 and Figure 8.13(b).

In practice, low pressure centers are also the signature of cyclones. Therefore we also say that the rotation of the winds around a low is **cyclonic**. By extension, the rotation of the winds around a high (an anticyclone) is said to be **anticyclonic**. Thus, in

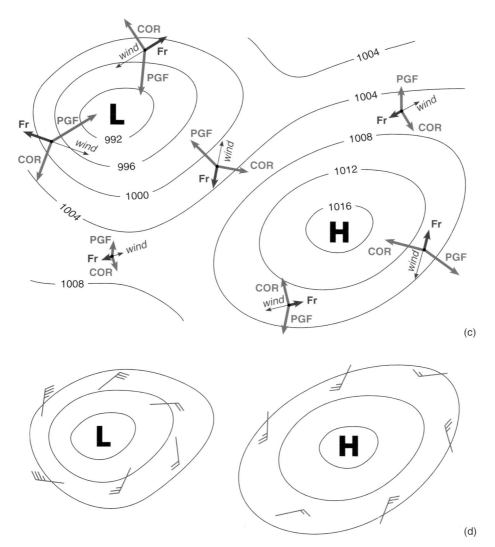

Figure 8.13. (cont.). (c) Balance of forces and wind pattern at the surface when taking friction (Fr) into account. (d) Convergent cyclonic circulation around a low vs. divergent anticyclonic motion around a high.

Box 8.5. Highs and Lows

Rule #5

Winds tend to blow clockwise around highs and counterclockwise around lows (in the northern hemisphere).

the northern hemisphere, cyclonic rotation is counterclockwise, while anticyclonic rotation is clockwise. In the southern hemisphere, flow is reversed, with cyclonic flow clockwise and anticyclonic flow counterclockwise. In both hemispheres, cyclonic flow is in the same direction as the local rotation of Earth when viewed from above the pole (Earth appears to rotate counterclockwise when viewed from above the North Pole, and clockwise when viewed from above the South Pole).

This simple rule is useful to form a general sense of the wind circulation, but a simple look at a surface weather map will reveal that the wind rarely blows exactly along the isobars. More often than not, it crosses the isobars at an angle that can be 45° and sometimes more. The deflection is due to friction between air and Earth's surface. Surface friction will indeed add an interesting twist to our picture.

8.8 Friction

When a moving object rubs against a surface, it experiences resistance in the direction of motion. The object could be our hand rubbing against a table, a ski sliding on a blanket of snow, water flowing on a river bed, or air blowing over a surface, be it ground, water, or a forest. This resistance, **friction**, is a force acting upon the object in the direction opposite the direction of motion, which slows the speed of the object. Thus, if no other force but friction is acting on a ski in motion, the ski will slow down and eventually come to a stop. Similarly, **surface friction** acts to slow down air in motion close to the surface, and the wind would stop completely if no other force was at work.

When surface friction acts on the wind, the wind speed is reduced. Therefore, the Coriolis force is also reduced, since it is proportional to wind speed. In this case the pressure gradient force is balanced by the sum of the friction and Coriolis forces, so that the wind must blow at an angle to the isobars, toward low pressure, as shown in Figure 8.14. The greater the friction effect, the more the wind speed will be reduced, and the greater the angle at which the wind will blow across the isobars.

The impact of surface friction is most pronounced at the surface and decreases with height, until it is very small in the free troposphere (above about 850 hPa). The lower layer of the troposphere where surface friction alters the force balance is called the **planetary boundary layer**. It can be very shallow (down to a few dozens of meters in some extreme circumstances, such as over the Antarctic ice sheet) or very deep, more than two kilometers in some hot and dry situations where convection constantly stirs the lower troposphere.

The surface type also plays a large role in determining the effects of friction at the surface. Some land types are very *rough*, such as forests and urban locations, while others are much smoother, such as grassland and deserts. Oceans are definitely smoother than land and, as a result, generate less friction. As a rule of thumb, winds tend to be deflected (relative to the isobars) by an angle of 15° to 30° over oceans, as compared to 30° to 45° over land.

Rules of motion for surface winds are summarized in Box 8.6 and illustrated in Figure 8.15, where

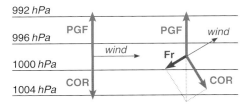

Figure 8.14. Comparison of the geostrophic approximation (left) and the new force balance including friction (Fr) close to the surface (right).

Box 8.6.   Surface Winds

Rule set #6

(1) Surface winds tend to blow at an angle to the isobars, pointing toward low pressure.

(2) Surface winds tend to converge into low pressure centers, and diverge away from high pressure centers.

(3) Convergence into lows is often associated with upward motion, cloud formation, and precipitation, while divergence out of highs is often associated with subsidence and clear skies.

---

surface winds have been overlaid on pressure and temperature. Note in particular how, in this view of our February 2014 cyclone, surface winds blow across the isobars and into the low, and outward of the high. This can be observed on surface weather maps every day, and will be an important part of understanding weather with surface cyclones (Chapter 10).

The frictional effect has very important consequences at the surface, as illustrated in Figure 8.13(c) on p. 137. Since surface winds tend to be deflected toward low pressure values, they will tend to be deflected *away from high pressure centers* and *toward low pressure centers*. In other words, winds tend to **diverge** away from highs, and to **converge** into lows (Figure 8.13(d), see p. 137). This is better illustrated with *streamlines*, as shown in Figure 8.16. Streamlines are drawn in such a way as to be everywhere parallel to the wind direction. They convey a sense of flow – in particular, nodes of convergence and divergence. Note in Figure 8.16 how the air diverges out of the

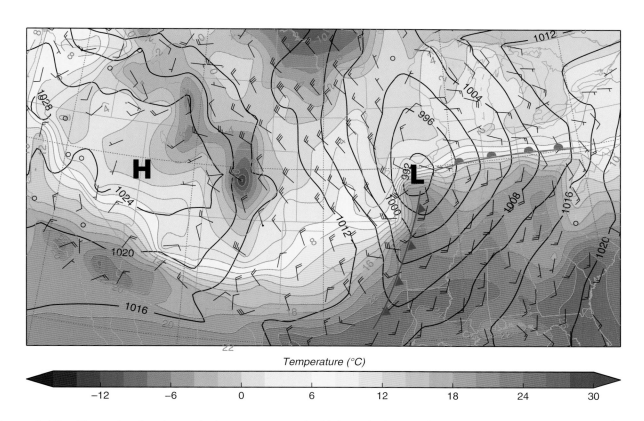

Temperature (°C)

−12   −6   0   6   12   18   24   30

Figure 8.15. Example of surface wind vectors around a high and a low on February 20, 2014, at 18:00 UTC.

high, flows southeastward toward the cold front, takes a sharp turn at the cold front to aim northward, and takes another sharp turn at the warm front, to continue westward and eventually into the low. Notice the many streamlines starting in the high pressure region and diverging away, as well as the many streamlines converging and ending into the low. [Note that streamlines are different from **trajectories**, which define the path of air over a period of time. Streamlines provide a snapshot of motion at one instant in time.]

These patterns of converging and diverging air have critical implications for the mass balance of weather systems. Converging air does not accumulate into a surface low, but continues on in a new direction. At the surface, since the ground is rigid, that new direction is upward, which provides a lifting mechanism that can enhance cloud formation and precipitation. That is one reason why low pressure centers (i.e., cyclones) are often associated with clouds and precipitation. Conversely, air diverging away from a surface high is replaced by air subsiding from aloft. Since subsidence inhibits cloud formation, high pressure regions (i.e.,

anticyclones) are often associated with few clouds or even completely clear skies.

Finally, you will also note in Figure 8.15 that winds tend to be stronger around lows and weaker around highs (and, correspondingly, isobars tend to be more closely packed around lows and more loosely spaced around highs). Indeed, converging winds swirling around a low tend to accelerate. Somewhat similarly to an ice skater spinning faster and faster as she pulls her arms and legs in toward the center line of her body, the rotating winds spin faster and faster around the low as they converge closer to the center of rotation. As a result, winds can be very strong near the center of deeper cyclones. Conversely, winds tend to slow down as they diverge away from a high pressure center, in the same way the ice skater slows down by spreading out her arms before stopping. That is why winds are usually very mild under an anticyclone.

With these general rules in mind, we are about ready to study all wind patterns around Earth. Before we do so, however, let us investigate a few additional factors that can influence wind development.

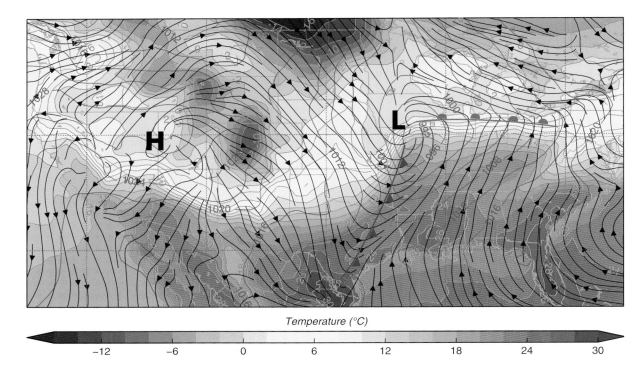

Temperature (°C)

−12   −6   0   6   12   18   24   30

Figure 8.16. Example of surface streamlines diverging out of a high and converging into a low on February 20, 2014, at 18:00 UTC.

8.9 Topography

The greatest influence on wind, after Earth's rotation, curvature, and friction, is that of topography. We have already mentioned the influence of orographic lifting on cloud formation, precipitation, and vegetation, as air is lifted by mountain slopes, as well as the impact of subsidence and adiabatic warming in the lee of mountains, with the formation of **chinook** winds, for example (see Chapter 6). Mountains of significant height can greatly alter the direction and speed of the wind at the surface, of course, but also in the higher atmosphere, as the deflection can be transmitted quite high above the mountain top. Mountain barriers such as the Andes, the Himalayas, and the Rocky Mountains can effectively force the air upward, or channel it at a completely different angle from that suggested by geostrophic balance. In the example shown in Figure 8.17, westerly flow from the Pacific Ocean is deflected by the Olympic Mountains and channeled through the Strait of Juan de Fuca (between the Olympic Mountains and Vancouver Island), and then is blocked and deflected by the Cascades. Note in particular how the winds are southerly in Seattle, Washington, where geostrophic balance would suggest westerly winds.

8.9.1 Mountain Breeze and Valley Breeze

Mountains and valleys can also induce local wind circulations through differential heating and buoyancy. During the day, the mountain slopes that form the "walls" of a valley heat faster than the air over the valley at the same altitude. Thus, the air parcels closest to the slopes are heated by conduction and are warmer than the environment. They are upwardly buoyant and tend to rise (Figure 8.18(a)). Relatively colder air sinks down into the valley and replaces the rising air along the slope. This causes a local circulation by which air tends to rise up the valley and flow along the mountain slopes, creating a valley breeze (blowing *from* the valley) that is compensated by subsidence down into the valley floor (Figure 8.18(b)). Such a valley breeze is usually quite mild, but it can lead to the formation of cumulus clouds above the top of the mountains and sometimes showers and afternoon thunderstorms.

The reverse circulation takes place after sunset. Emission of infrared radiation by the mountains cools the valley walls, which in turn cool the adjacent air. Becoming colder than air at the same elevation over the valley, the air is negatively buoyant and sinks down

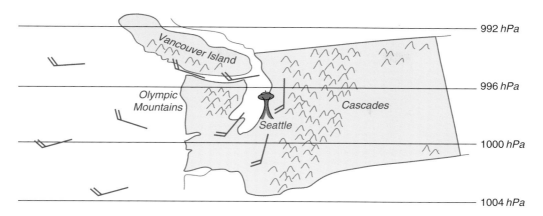

Figure 8.17. Blocking and channeling of westerly winds by the Olympic Mountains, upstream of Seattle.

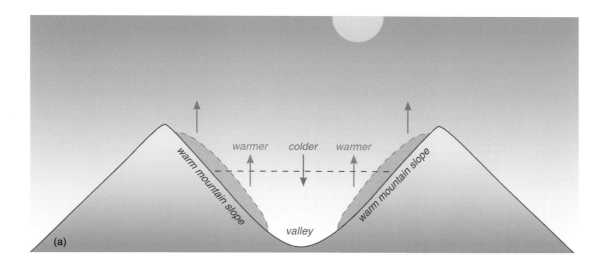

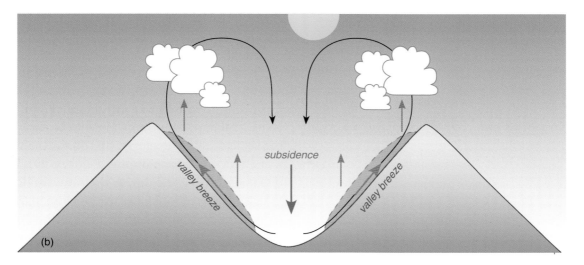

Figure 8.18. Development of a valley breeze during daytime.

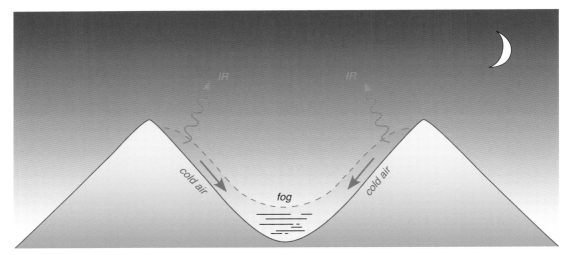

Figure 8.19. Cooling by emission of infrared radiation (IR), downslope flow of cold air at night, and formation of fog in the valley when the temperature drops below the dew point.

Figure 8.20. Valley fog in the Central Valley of California.

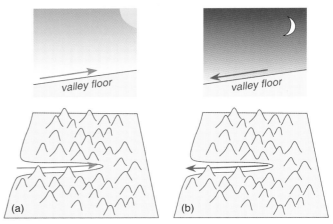

Figure 8.21. Daytime up-valley flow (a) vs. nighttime down-valley flow (b).

the mountain slopes into the valley. Being denser than the warm air above, the cold air flows down the valley, like the rivers that are also found there. In places where the valley floor is relatively flat, cold air can accumulate, cooling even further as the night progresses. If the dew point is reached, condensation takes place and **valley fog** forms, trapped in the valley by the temperature inversion (Figure 8.19). Valley fog is very common in autumn, and can be easily observed in visible satellite imagery, as it follows the valleys and meanders of the topography (Figure 8.20). It typically dissipates during the day, as the sun warms the cloud layer and convection mixes the air close to the surface.

As a result of this diurnal cycle of heating, in the base of the valley, it is common to observe a breeze blowing up-valley in the afternoon and down-valley at night (Figure 8.21).

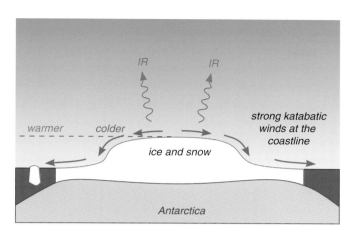

Figure 8.22. Schematic representation of very cold air accelerating downslope over Antarctica to produce strong katabatic winds at the coastline. (The proportions have been greatly exaggerated.)

8.9.2 Katabatic Winds

When very cold and dense air flows down especially high topography, it can give rise to some of the strongest and persistent winds on Earth's surface. This is the case in Greenland, and especially Antarctica, where cold air drains off an elevated continental mass covered with ice and snow. The combination of extremely low temperatures, due to radiational cooling and the absence of sunlight in winter, results in the development of cold high pressure systems. Air flows away from these high pressure systems, diverging and accelerating away from the continent, in very strong winds called **katabatic winds**. The air flows downslope toward the edge of the ice-covered continental mass because it is very dense and therefore negatively buoyant, accelerating under the effect of gravity to reach a maximum wind speed at the coastline (Figure 8.22). Even though the air warms up by adiabatic compression, it remains colder than the environmental air it displaces.

These cold winds exiting the continental mass continue northward and feed into the cold sector of midlatitude cyclones developing over the Southern Ocean. Indeed, winds are not isolated phenomena. They are all connected within the larger system that is the dynamic atmosphere. Thus, understanding local winds also means understanding the larger

atmospheric circulation at the scale of the planet, which will be the focus of Chapter 9.

Summary

Horizontal motion of air is largely determined by the **pressure gradient force** and the **Coriolis force**, with the addition of the centrifugal force in regions of curved flow, and friction close to the surface.

The **pressure gradient force** results from horizontal variations in atmospheric pressure, and, in the absence of other forces, accelerates the wind from high to low pressure.

In the absence of weather systems, a **sea breeze** (wind blowing from sea to land) will form at a coastline during the day when air over land becomes warmer than the air over the adjacent water surface. At night, when air over land becomes colder than air over water, a **land breeze** (wind blowing from land to sea) forms.

Earth's rotation creates an apparent force (the **Coriolis force**) that tends to deflect the air at a right angle and to the right of the direction of motion in the northern hemisphere (and to the left in the southern hemisphere). The Coriolis force is zero at the equator, and increases with latitude. It is proportional to the wind speed, but only affects the wind direction (i.e., it does not affect the wind speed).

When the pressure gradient force and the Coriolis force are equal and opposite, resulting in no net force, the wind is said to be in **geostrophic balance**. The wind then blows parallel to the isobars with higher pressures on the right in the northern hemisphere (on the left in the southern hemisphere), and the wind speed is proportional to the magnitude of the pressure gradient.

When the flow is curved, the **centrifugal force** makes the wind **subgeostrophic** around a trough and **supergeostrophic** around a ridge.

At the surface, winds tend to blow clockwise around highs and counterclockwise around lows in the northern hemisphere (opposite in the southern hemisphere), but **friction** slows the wind speed and changes the direction so that the wind blows at an angle to the isobars, toward low pressure. Therefore, surface winds tend to **converge** into low pressure centers, causing upward motion, cloud formation, and precipitation, and **diverge** away from high pressure centers, causing subsidence and clear skies.

Topography can deflect and channel the wind at a different angle and speed than those suggested by geostrophic balance (e.g., mountain ranges, passes, straits).

Differential heating along valley slopes can cause local circulations by which air tends to rise and flow up the mountain slopes during the day, creating a **valley breeze** with afternoon clouds and showers over the mountains, and to sink down the mountain slopes at night with the formation of **valley fog** if the dew point is reached.

Katabatic winds form when very cold and dense air flows down an elevated plateau, accelerating down the slopes under the effect of buoyancy and, often, additional factors (e.g., subsidence, pressure gradient), as is typical in Greenland and Antarctica.

CHAPTER 9

Global Wind Systems

Since they impact our daily life, we tend to think about wind and weather locally, but they are in fact interconnected with the dynamics of heat transfer in the atmosphere. We will now step back and apply our concepts of differential heating, pressure gradient forces, and geostrophic motion at a global scale, to explore the implications for what is called the general circulation of the atmosphere, which will set the broader context within which individual weather systems exist.

CONTENTS

Weather is ever changing; however, some features persist through this daily variability and guide the seasonal rhythms of life. Subsaharan countries such as Ghana and Burkina Faso, for example, can rely on a yearly rainy season from about April to November. By contrast, high pressure over the Sahara is consistent all year round and prevents cloud formation and precipitation, maintaining a desert. In midlatitudes, the presence of a high pressure system over the North Pacific Ocean insures consistently mild weather on the West Coast of the United States in the summer. This suggests that there are somewhat permanent ("semi-permanent") features that anchor the weather. Although daily variations can depart substantially from these average features, they vary around this semi-permanent state, which provides some level of consistency.

To isolate the semi-permanent state, we first start by **averaging** atmospheric data over long periods of time. This procedure erases the detail of individual weather events, but reveals persistent patterns that we call the **general circulation** of the atmosphere. Then we elaborate a model that accounts for the observed averaged fields and provides a framework for understanding variability in weather and climate events.

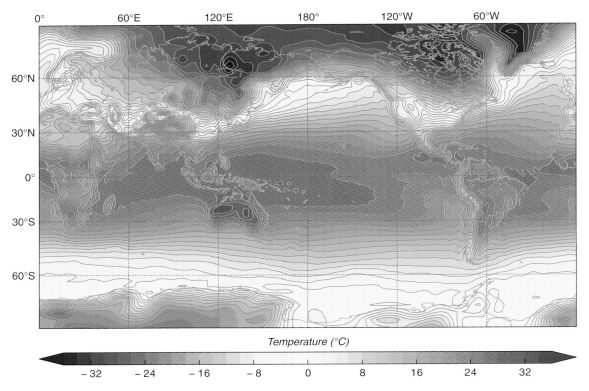

Figure 9.1. (a) Mean surface air temperature for the month of January.

9.1 The Averaged Atmosphere

9.1.1 Surface Temperature

Consider the average distribution of air temperature at the surface shown in Figure 9.1. These maps were created by averaging surface air temperature over 68 years for January and July, respectively. We recognize familiar features, such as the warm equatorial region and the cold polar regions, which we know exist because of the differential heating of Earth and the resulting accumulation of energy at low latitudes compared to high latitudes (see Chapter 4). Figure 9.1 also shows that, whereas the isotherms are relatively widely spaced at low and (to a lesser extent) high latitudes, the region of transition between the warmer tropics and the colder polar regions, i.e., the midlatitudes, corresponds to a region of strong temperature gradient. This is most obvious in the southern hemisphere in winter (Figure 9.1(b)), because of the presence of Antarctica, where the coldest temperatures on Earth's surface are recorded. The temperature

contrast between the extremely cold continent and the warmer ocean creates a very strong temperature gradient around 60 °S, reminiscent of a temperature front. A similar pattern is found in the northern hemisphere in winter (Figure 9.1(a)) between 30 °N and 60 °N. The presence of cold land masses complicates the picture in that the gradient meanders south over land areas and north over the oceans, but the sharp temperature gradient is clearly visible, both in the color shading and the tightness of the isotherms. This transitional region is sometimes called the **polar front**.

Note that the temperature gradient is much weaker in the southern hemisphere in January (Figure 9.1(a)) and in the Northern Hemisphere in July (Figure 9.1(b)); i.e., in the summer hemisphere. Conversely, the temperature gradient is strongest in the winter hemisphere. Figure 9.2 shows schematically why this should be so. The tropics, being more directly under the sun, are relatively warm independent of the season. The polar region, however, becomes extremely cold in winter, because it is pointing away from the sun and receives little sunlight

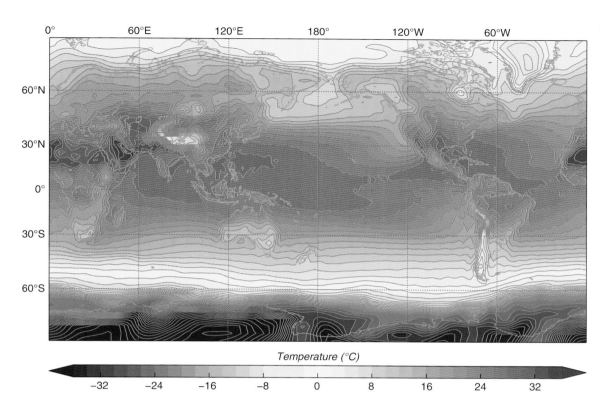

Figure 9.1. (cont.). (b) Mean surface air temperature for the month of July.

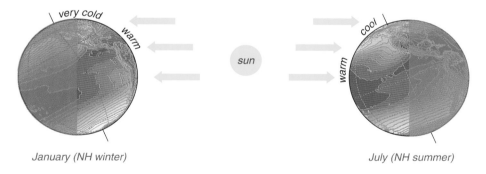

January (NH winter)

July (NH summer)

Figure 9.2. The temperature contrast between the tropics and the polar region is much stronger in winter than in summer.

during few hours of daylight, if any, but is constantly losing heat to space by emission of infrared radiation. Consequently, the temperature contrast between the tropics and the poles is much greater in winter than in summer.

Figure 9.2 also reminds us that the region most exposed to the sun shifts north and south over time. It shifts further south in January and further north in July. We can verify in Figure 9.1 that the band of highest

temperatures indeed shifts to the south in January (see Australia in Figure 9.1(a), for example) and to the north in July (see northern Africa in Figure 9.1(b)). This will have important implications later on when we add pressure and winds to the picture.

The surface temperature distribution has direct consequences on the pressure distribution around Earth, as we will now explore by first considering the upper troposphere.

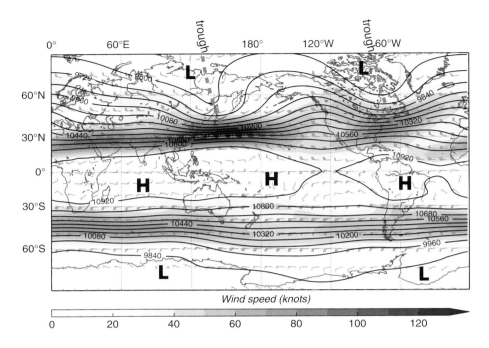

Figure 9.3.   (a) Average 250 hPa heights, winds, and wind speeds for the month of January with schematic representation of higher and lower pressures, and upper-level troughs. Areas of maximum wind speed indicate the position of the jet streams.

9.1.2 Upper-Level Heights

First, a map of average 250 hPa heights reveals a broad swath of higher heights over the equator (recall the equivalence of height values on pressure surfaces with pressure values on height surfaces), a large area of lower heights anchored around Antarctica, as well as lower pressures at the North Pole, although the latter are less intense in the summer (Figure 9.3). In the northern hemisphere, two prominent troughs are found over Eastern Asia and the eastern part of North America in winter (Figure 9.3(a)). They are due to the presence of the two continental land masses, affecting air flow through land–sea temperature contrasts as well as topography. These upper-level troughs will play an important role in explaining the concentration of midlatitude cyclones in those regions (see Chapter 10).

We also note in Figure 9.3 that the upper-level winds generally blow from west to east (i.e., westerly winds). We can now explain this observation if we recall that, at the planetary scale, winds are influenced by Earth's rotation and adjust to geostrophic balance. Therefore,

winds should blow parallel to the isobars (or height contours) with higher pressures (heights) on their right and lower pressures (heights) on their left in the northern hemisphere, and conversely in the southern hemisphere, which is indeed what we observe. Most interesting, however, is the fact that these westerlies are concentrated in narrow regions of very high wind speed in both hemispheres, where the meridional pressure gradient is most intense, as indicated by the tightness of the height contours around 30–60 °N and 30–60 °S, i.e., in the midlatitudes. The blue ribbons of Figure 9.3 are indeed the strong upper-level jets that we identify as the **jet streams**, as shown in Figure 9.3(c). They reach wind speeds higher than 100 knots in these average maps, and much higher wind speeds are found on individual days. Moreover, we can verify that the jet streams reach their maximum wind speed not only in the midlatitudes, but also at the tropopause, i.e., at about 250 hPa (see Figure 5.9 for a nice illustration). The northern hemisphere and southern hemisphere jet streams are shown schematically in Figure 9.4(a), where the color-shaded wind speeds of Figure 9.3(a) have been retained in the

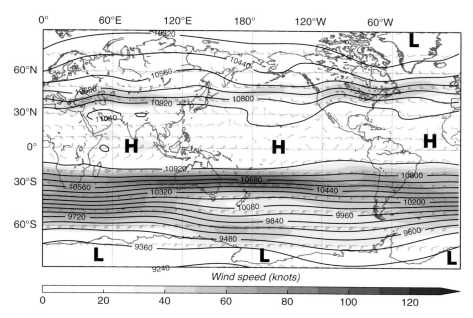

Figure 9.3. (cont.). (b) Average 250 hPa heights, winds, and wind speeds for the month of July, with schematic representation of higher and lower pressures. Areas of maximum wind speed indicate the position of the jet streams.

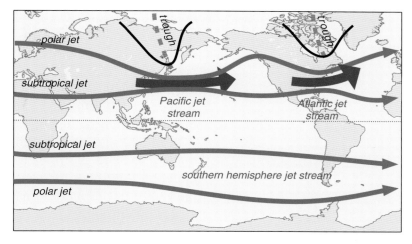

Figure 9.3. (cont.). (c) Schematic representation of the semi-permanent jet streams and upper-level troughs (in winter for each hemisphere).

background. The jet streams are shown both as blue arrows indicating westerly motion, as if viewed from above, and as blue crosses representing the tail of the arrows going into the page, as if we were looking at the arrows from behind.

To explain the presence and the intensity of the jet streams, we need to return to our average temperature maps (Figure 9.1) and recall that there exists a strong temperature gradient at the surface, in the midlatitudes (Figure 9.4(b)). If we lay out this temperature gradient on a flat surface, as if we were flattening Earth's surface, or as if we were looking at Figure 9.1(a) from the west, we obtain the schematics shown in Figure 9.4(c). We can now add a cross-section of the atmosphere and show a few pressure surfaces. Recall that, where the atmosphere is warmer (i.e., in

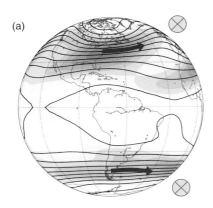

(a)

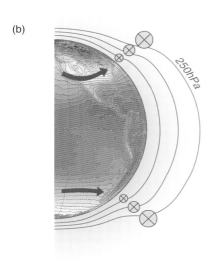

(b)

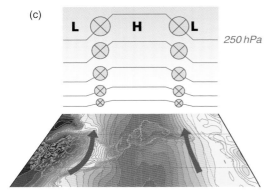

(c)

Figure 9.4. (a) Schematic representation of the jet streams, both as a lateral view (blue arrows) and a rear-end view, as if looking at the tail of the arrows going into the page (blue crosses). The blue shading indicates wind speed as in Figure 9.3. (b) Increase in westerly wind speed with height in the midlatitudes shown on three pressure surfaces. The jet stream reaches maximum wind speed at the tropopause (250 hPa). (c) Same as (b) after flattening Earth's surface.

the tropics), the atmosphere is "thicker" and the pressure surfaces are raised compared to the poles. If we now compare two points at the same height near the tropopause, we find that the pressure is higher at the top of the warm air column (i.e., above the tropics) and lower at the top of the colder air column (i.e., over the poles). As shown in Figure 9.4(c), and in accordance with geostrophic balance, the resulting winds are westerly, and into the page.

But Figure 9.4(c) also shows that the slope of the pressure surfaces over the midlatitude temperature gradient increases with height, i.e., the pressure surfaces are increasingly steeper with height. In other words, the pressure gradient increases with height and, according to geostrophic balance, the wind speed increases. It turns out that it increases in such a manner up to about 250 hPa. Recall that, above the tropopause, the temperature increases with height due to warming from ozone absorption of ultraviolet radiation. Since we cross the tropopause over polar regions at lower altitudes than over tropical regions, the temperature decrease with latitude becomes smaller with height above the tropopause and eventually switches direction. As a result, the temperature gradient, and hence the pressure gradient, decrease above about 250 hPa. Therefore, the wind speed decreases as well, and we are left with a jet of maximum wind speed in the middle, right at the tropopause; i.e., the jet stream.

In summary, the temperature distribution is such that there exists a strong temperature gradient in midlatitudes, which induces a pressure gradient in the upper troposphere, which itself induces a westerly jet. This allows us to explain yet another subtlety of Figure 9.3. You might have noticed that the jet stream is strongest in January in the northern hemisphere, and strongest in July in the southern hemisphere. In other words, the jet stream is always stronger in the winter hemisphere. By now this should not come as a surprise, since we know that the jet stream is intimately tied to the midlatitude temperature gradient. Indeed, recall from Figures 9.1 and 9.2 that the midlatitude temperature gradient is always stronger in the winter hemisphere. It stands to reason that a stronger temperature gradient should induce a stronger pressure gradient and a stronger jet stream. (The seasonal contrast is less evident in the southern hemisphere than in the northern hemisphere because of the relative absence of continents relative to open water and, as a consequence of the thermal inertia of water discussed in Chapter 4, less seasonality.)

Theory tells us that, given the size and rotation rate of Earth, it is on the borderline of supporting two distinct jet streams. (Jupiter, for example, is huge compared to Earth, with much faster rotation, and has many jet streams). Indeed, if we look carefully, we find that the jet streams meander and sometimes split. In winter, in particular, we can guess the presence of two separate jets entering North America and two separate jets entering Europe (Figure 9.3(a)), as well as two separate jets over the South Pacific Ocean (Figure 9.3(b)). Indeed, there are often two jets, a polar jet and a subtropical jet, which sometimes combine to form a stronger, unique jet in some areas (Figure 9.3(c)): in particular, a Pacific jet stream over the North Pacific Ocean, east and downstream of Japan; and an Atlantic jet stream over the North Atlantic Ocean, east and downstream of New England. Recall that these features are revealed after averaging a large number of daily situations, and are therefore what is found on average. As such, these jet streams can be expected to exert a strong constraint on the average weather. These two regions where the jet stream is strongest are found south of the persistent upper-level troughs extending south from Eastern Siberia and Eastern Canada, and correspond to areas of enhanced midlatitude cyclogenesis. We will see in Chapter 10 that local meanders of the jet stream, short-wave troughs, are critical for the development of midlatitude cyclones.

9.1.3 Surface Pressure

At the surface, the pressure field is broken up in closed highs and lows that more or less follow the distribution of land masses and oceans (Figure 9.5). In particular, we note the presence of **subpolar lows** at about 60–70 °N and 60–70 °S, especially in winter, as well as the presence of **subtropical highs** at about 30 °N and 30 °S. The subtropical highs tend to form on the eastern side of the ocean basins. Note that, in the southern hemisphere, amidst the highs and lows, there persists a continuous stream of midlatitude

westerlies at about 40–50°S, where the ocean is continuous from west to east. These particularly strong winds were called the "roaring forties" by early sailors, who knew to sail from west to east to take advantage of the general direction of the winds, but to avoid the regions of strongest winds, as they are treacherous and can sink a ship.

The lows are stronger in winter, and we give them particular names when describing and forecasting the weather. In the northern hemisphere, the Aleutian Low is centered over the Aleutian Islands, a chain of islands extending southwest from mainland Alaska, while we usually find the Icelandic Low near Iceland (Figure 9.5(a)). Similarly, the subtropical highs are stronger in summer, and we usually refer to them as the Pacific High and the Bermuda/Azores High in the northern hemisphere, the latter being typically found between Bermuda and the Azores Islands (Figure 9.5(b)).

9.1.4 Precipitation

A map of average precipitation also contains a number of interesting features (Figure 9.6). The equator features a band of heavy rain, while the polar regions show little precipitation. In the midlatitudes, between about 30° and 60° north and south of the equator, we recognize rain patterns near the jet streams over the Pacific and Atlantic Oceans corresponding to precipitation from midlatitude cyclones (see Chapter 10). We also notice strikingly dry areas in the subtropics, in particular over the eastern side of the ocean basins and some land areas, such as Australia and the Sahara desert. (Note the correspondence with the subtropical highs of Figure 9.5.)

Our goal is now to elaborate a sparse model of the circulation of the atmosphere that articulates all of these semi-permanent features. We start with the main feature of the energy distribution of the atmosphere: the temperature contrast between the equator and the poles. Recall that temperature contrasts create pressure gradients aloft, which set the atmosphere in motion. This was illustrated with the sea breeze circulation in Chapter 8. Even though the sea breeze mechanism applies only to local-scale circulations, the

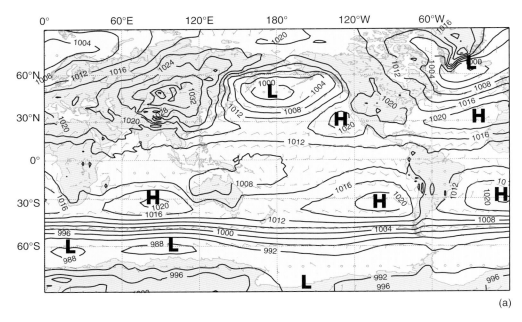

Figure 9.5. (a) Mean sea level pressure for the month of January.

physical discussion that underlies it constitutes a good starting point for building our first model of the general circulation of the atmosphere.

9.2 The Single-Cell Model

Recall that, in a sea breeze circulation, air over the warmer area flows toward the colder air aloft, increasing the surface atmospheric pressure over the colder area and decreasing it over the warmer area (Figure 9.7(a)). Air subsides over the colder area and the sea breeze returns from the cold to warm area at the surface, closing the circulation.

If we start by assuming that Earth does not rotate and we view the temperature difference between the equator and the poles as such a temperature contrast, but taking place over much larger distances (Figure 9.7(b)), we can expect the same response such that the air should rise over the equator and diverge away from the equatorial region aloft (at the tropopause, or about 250 hPa). As a result, the surface pressure should drop at the equator, creating a trough of lower pressures. As upper-level air converges at the poles, the surface pressure should increase and the air should sink, closing this double

circulation (northern and southern hemispheres). Thus we should expect poleward upper-level winds and equatorward surface winds.

This circulation, named after George Hadley (1685–1758), who first proposed it, is called a **Hadley cell**, and our simple model suggests that there should be a Hadley cell in each hemisphere (Figure 9.8). This picture is indeed consistent with aspects of our earlier observations of the mean atmosphere. At the tropopause, we indeed observe higher pressures over the equator, and lower pressures at the poles (Figure 9.3). The Hadley cell model depicts converging winds at the equator, which is also consistent with observations. Indeed, note in Figure 9.5 how, at low latitudes, the surface winds tend to have a northeasterly direction north of the equator, and a southeasterly direction south of the equator. These winds, called the **trade winds**, necessarily meet and converge in the equatorial region, to form a semi-permanent area of convergence called the **Intertropical Convergence Zone** (ITCZ). Since converging winds tend to rise (see Chapter 6), the ITCZ is coincident with convection and precipitation, as confirmed by the band of higher rain rate visible in the equatorial region in Figure 9.6, and by the cumulonimbus clouds often observed in satellite imagery (Figure 9.9). This band of heavy

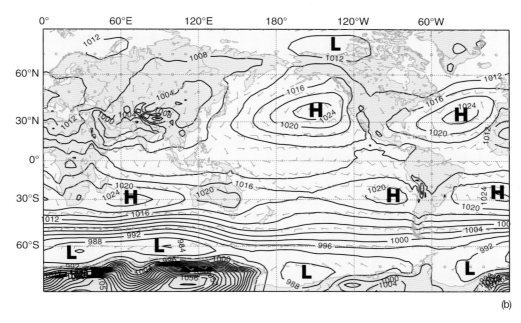

Figure 9.5. (cont.).   (b) Mean sea level pressure for the month of July.

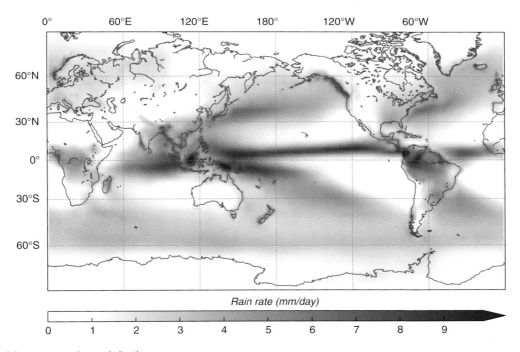

Rain rate (mm/day)

Figure 9.6.   Mean annual precipitation.

precipitation observed at the equator is consistent with equatorial convection predicted by the Hadley cell model (refer to Figure 9.8), while reduced precipitation in the polar regions is consistent with subsidence. Indeed, low pressures and convection at the equator should produce clouds and precipitation, while subsidence at the poles should inhibit cloud formation and precipitation.

The bands of precipitation observed in the midlatitudes, however, are not explained by the simple single-cell Hadley model. We also noted in Figure 9.3 that the upper-level winds are not blowing from high

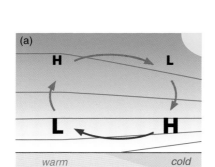
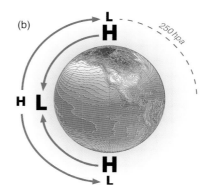

Figure 9.7 (a) Schematic representation of a sea breeze created by a horizontal temperature contrast. (b) Sea breeze concept applied to the temperature contrast between the equator and the polar regions.

to low pressures in the midlatitudes as our two Hadley cells would suggest (i.e., in a north–south direction), but overall from west to east. Similarly, whereas the Hadley cell model suggests the presence of only low surface pressures at the equator and high surface pressures at the poles, we noted that, in reality, the surface pressure field is broken up in closed highs and lows in midlatitudes.

In summary, our Hadley cell model has some qualities consistent with observations, such as convergence and precipitation near the equator, and subsidence at the poles. However, it does not have westerly jet streams or highs and lows in the subtropics and midlatitudes, which require a more detailed model.

9.3 The Three-Cell Model

Our single-cell model neglects one crucial component: Earth's rotation. Indeed, as the upper-level winds blow away from the equator and poleward, they tend to be deflected to the right in the northern hemisphere and to the left in the southern hemisphere (i.e., eastward in both cases) due to the Coriolis acceleration (see Chapter 8). Similarly, surface winds blowing toward the equator tend to be deflected to the right in the northern hemisphere and to the left in the southern hemisphere: westward in both cases (Figure 9.10). In other words, this should result in upper-level westerlies and surface easterlies.

It turns out that this configuration is unstable, and the circulation breaks down into a pattern that is better captured by a three-cell model, as shown in Figure 9.11. The upper-level flow of the Hadley cell returns to the surface in the tropics, between 20 and 30 °N in the northern hemisphere, and between 20 and 30 °S in the southern hemisphere, which limits the extent of the **Hadley cells** to about 30° from the equator. Similarly, the rest of the single cell breaks down into a **polar cell** spanning about 60–90° latitude and an intermediate region spanning 30–60° latitude. We refer to this midlatitude region as the **Ferrel cell**, after William Ferrel (1817–1891), although the circulation there does not really take the form of a cell. While the Hadley and polar cells are somewhat representative of the real circulation and are conceptually useful, midlatitude dynamics are largely defined by the development and propagation of extratropical cyclones (see Chapter 10), for which the circulation is far more complex. The Ferrel cell represents the average effect of these features, and is therefore more of a mathematical artifact that depends on how the data is averaged. (For example, if, rather than latitude, we use temperature as a north–south coordinate for averaging, the Ferrel cell disappears.)

Besides capturing the features previously described (ITCZ, convection at the equator, subsidence at the poles), the three-cell model also suggests the presence of subsidence in the subtropics, between 20°N/S and 30°N/S, which explains the presence of the subtropical highs observed in Figure 9.5, as well as the dry areas observed in Figure 9.6. (Recall that subsidence prevents cloud formation, and therefore precipitation.) In fact, this latitude band contains most of the major deserts on Earth (Figure 9.12). (Also note the location of the rain forests in this figure, consistent with convection and precipitation at the ITCZ.)

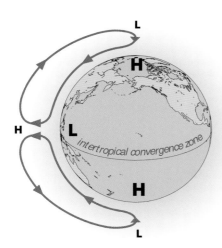

Figure 9.8. Single-cell model.

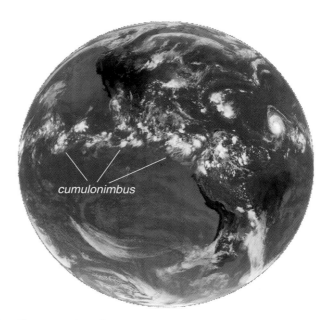

Figure 9.9. Cold cumulonimbus cloud tops indicate the location of the Intertropical Convergence Zone (ITCZ) in this (GOES-14) infrared satellite image.

Combined with Earth's rotation and the Coriolis acceleration, the three-cell model also explains the direction of the trade winds converging into the ITCZ. These winds are northeasterly in the northern hemisphere and southeasterly in the southern hemisphere, consistently with the deflection of the surface branch of the Hadley cell (recall Figure 9.10) and with the direction of the circulation on the equatorward flank of the subtropical highs.

In the end, the three-cell model constitutes a useful framework for understanding general weather and climate features at the regional scale, as we will now illustrate with some examples.

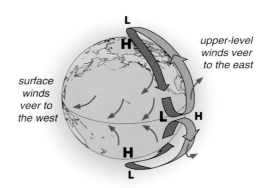

Figure 9.10. Distortion of the Hadley cell due to Earth's rotation.

9.4 Some Large-Scale Circulations

9.4.1 West Coast vs. East Coast

Summer weather on the West Coast of the United States is in stark contrast with that on the East Coast. While Seattle, Wasington, Portland, Oregon and San Francisco, California are experiencing relatively cool summer temperatures and dry weather, Georgia and the Carolinas are experiencing warm and moist conditions, with frequent convective summer rain. These sharp differences can be explained by considering the average surface pressure distribution shown in Figure 9.5 and reproduced in Figure 9.13

with schematic winds. While the Pacific high produces northwesterly to northerly winds down the West Coast, bringing cool, dry air from the Pacific Ocean and Canada, the Bermuda High produces southerly winds, bringing heat and moisture from the tropics toward Florida and the East Coast. Therefore, although at similar latitudes, and both in a coastal environment with apparently similar exposure to the ocean, the two regions experience very different summer weather conditions.

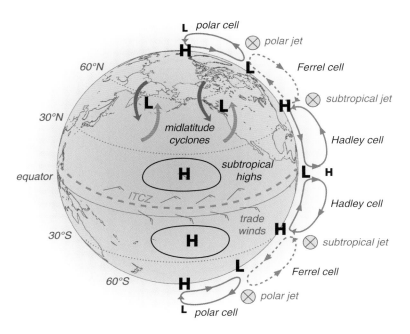

Figure 9.11. Three-cell model.

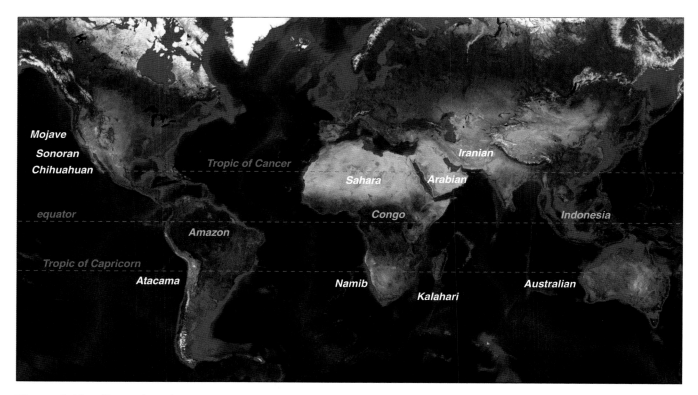

Figure 9.12. Examples of tropical rain forests along the equator (in green) and subtropical deserts along the tropics (in white).

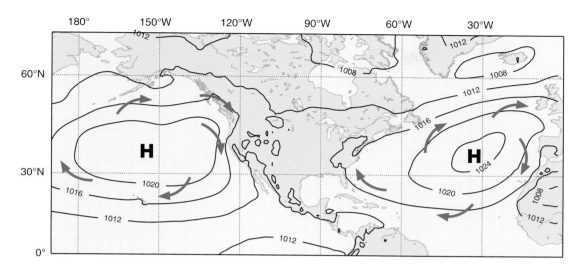

Figure 9.13.   Summer prevailing winds on the West Coast and East Coast of the United States, as determined by the semi-permanent highs over the Pacific Ocean and Atlantic Ocean.

9.4.2 Antarctica

We can apply similar reasoning to the southern hemisphere. Recalling that there is subsidence and high pressure over the South Pole (see Figure 9.8), we can explain why surface winds diverge toward the surrounding Southern Ocean (Figure 9.14). These surface winds are the **katabatic winds** encountered in Chapter 8. Recall that the Antarctic ice cap is so cold that the overlying air, cooled by conduction, sinks close to the surface and flows downslope toward the ocean, channeled by the topography of the glaciers. All these effects (high pressure, subsidence, divergence, higher density due to temperature, downslope flow due to negative buoyancy) combine to create strong, cold air currents that exit the continent and surrounding ice to continue their course northward.

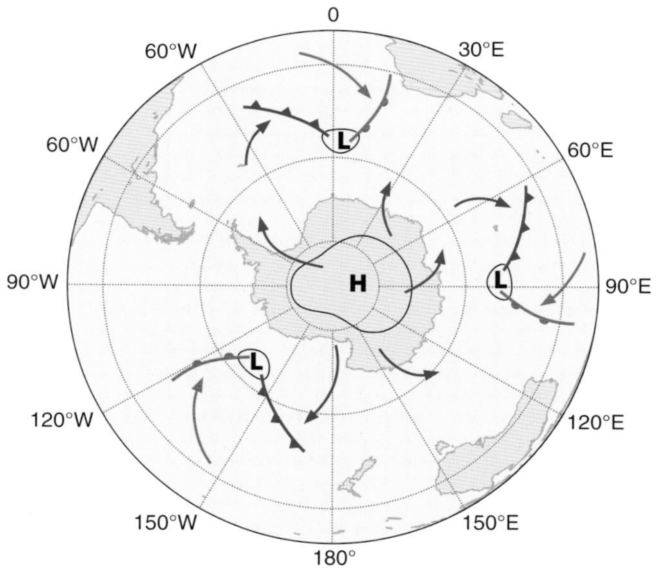

9.4.3 The Sahel

The general circulation is of course not static. It is an average picture that gives us a general idea of what to expect. On short time scales, weather systems respond to other forcings that create the variability we observe. The primary variation we experience throughout the year is due to the seasonal change in solar insolation.

Figure 9.14.   Katabatic winds and midlatitude cyclones resulting from the general circulation of the southern hemisphere.

Figure 9.3 illustrates the impact of this seasonal variation on the upper-level pressure distribution and jet streams, while Figure 9.5 illustrates its impact on surface pressure patterns and winds.

In the tropics, this seasonal variation directly impacts the position of the Intertropical Convergence

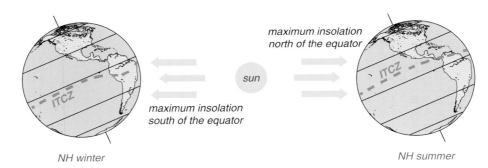

Figure 9.15. The ITCZ is displaced southward during (northern hemisphere) winter and northward during (northern hemisphere) summer.

Zone (ITCZ) and associated weather. Recall that the ITCZ is an area of wind convergence, associated with the formation of cumulonimbus clouds and heavy precipitation. Convection is driven at least in part by the absorption of incoming solar radiation by the underlying land or ocean. Therefore, the position of the ITCZ responds in part to the location of maximum insolation. As shown in Figure 9.15, this maximum shifts southward during northern hemisphere winter and northward during northern hemisphere summer. As a result, the ITCZ is displaced southward and northward accordingly. Note, however, that other factors control the ITCZ, such as the presence of land or ocean and the dynamics of the Hadley cells. Therefore, the displacement is not uniform around the globe, and it is not symmetric between the two hemispheres. In particular, the northward displacement is typically larger than the displacement into the southern hemisphere.

A good example of the impact of this seasonal variation of the ITCZ on local weather is found in Africa, in the Sahel, a narrow band of semi-arid grassland at about 15°N, between the equatorial rain forests and the Sahara desert (Figure 9.16(a)). In winter, the ITCZ is displaced southward, and northerly winds bring warm, dry air to the Sahel. Recall that the ITCZ is inherently a region of low pressure; therefore, winds tend to blow from the subtropical highs toward the ITCZ (Figure 9.16(b)). In summer, by contrast, the ITCZ is displaced northward, along with the associated convection and precipitation, as southerly winds bring moisture from the ocean (Figure 9.16(c)). As a result, the Sahel experiences a few rainy months that

bring enough water to sustain a steppe-like environment and a modest level of agriculture. Thus, while the equatorial rainforests experience a wet regime during a large fraction of the year, and the Sahara experiences a desert climate all year round, dry and wet seasons alternate in the Sahel every year.

Note that climate models predict that, in a warmer world such as projected due to global warming, desertification will threaten the Sahel and the local populations it sustains (see Chapter 15).

9.4.4 The Indian Monsoon

A similar displacement of the ITCZ can be observed in the Indian Ocean, and greatly affects the weather in India and surrounding countries. It is enhanced by two additional factors. First, the land–ocean temperature contrast drives a wind circulation similar to a sea breeze, except on a much larger spatial scale, and therefore subject to Earth's rotation and the Coriolis force. Second, the Himalaya mountains and the Tibetan Plateau to the north, which act as an elevated heat source or sink, enhance convection or subsidence at different times of the year (Figure 9.17(a)).

In Northern Hemisphere winter the ITCZ shifts southward and a high pressure region settles over the Tibetan Plateau (Figure 9.17(b)). The resulting northerly winds bring dry air and reduced precipitation to India from about November to May. Since these winds descend along the slopes of the Himalayas, the subsiding air warms, which makes for an often hot, dry season, especially toward April–May.

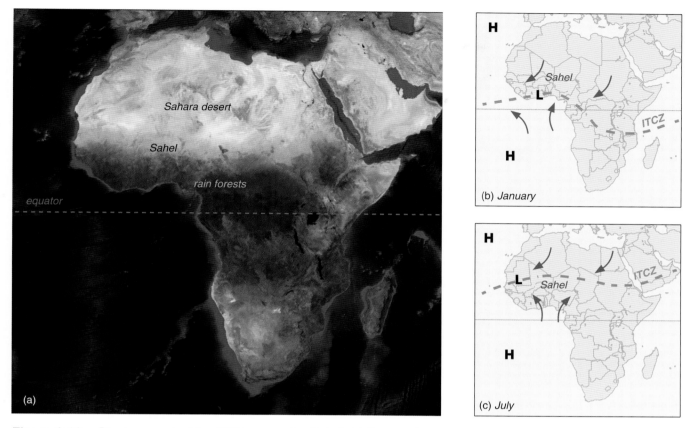

Figure 9.16. Displacement of the ITCZ across the Sahel. (a) Composite satellite view of Earth showing rain forests and desert areas in Africa. (b) Southward displacement of the ITCZ in northern winter. (c) Northward displacement of the ITCZ in northern summer.

In northern hemisphere summer the ITCZ shifts northward and, the continent being much warmer than the ocean, a continent-scale sea-breeze type of thermal circulation develops, which enhances the high-to-low surface winds from the Indian Ocean to the continent (Figure 9.17(c)). This enhancement moves the ITCZ much farther north than would normally be the case. Warm temperatures over land also enhance convection, and the southerly flow tends to push the air against the Himalayas, forcing it to rise. Since the southerlies originate over the warm Indian Ocean, they contain abundant water vapor, and convergence along the ITCZ, convection, and orographic lifting all combine to produce heavy precipitation during the rainy season, from about July to September. The eastern part of India is notorious for receiving record amounts of precipitation (as high as three meters) in short amounts of time.

9.4.5 El Niño

The best known fluctuation in tropical climate starts off the coast of South America, where the subtropical high pressure system drives southerly winds (Figure 9.18). As the wind blows over the ocean, it drags the surface water along with it. However, since water in the ocean is subject to the same Coriolis acceleration as air in the atmosphere, it tends to be deflected to the left (in the southern hemisphere), at a right angle to the wind, as shown schematically with gray arrows in Figure 9.18. As the surface waters move away from the coast, they are replaced by deeper water rising from the deep ocean. This pumping of deep, cold, nutrient-rich water is called **upwelling**. It is well known to local fishermen, as the nutrient-rich waters attract many fish.

Along the equator, a similar mechanism is at work. As the easterly trade winds converge toward the ITCZ,

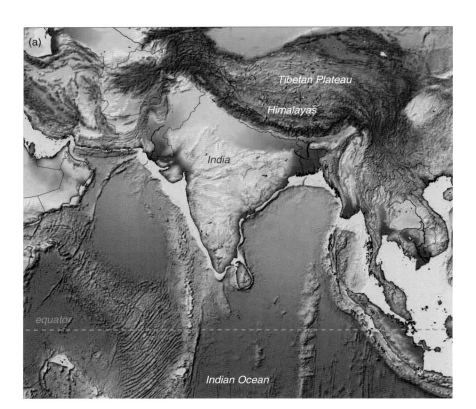

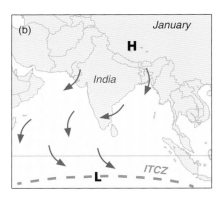

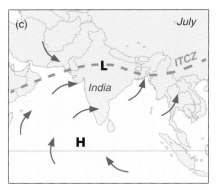

Figure 9.17. Displacement of the ITCZ during the Indian monsoon. (a) Topography of India and the Tibetan Plateau. (b) Southward displacement of the ITCZ in northern winter. (c) Northward displacement of the ITCZ in northern summer.

they also drag surface water with them. The Coriolis force deflects the water northwestward in the northern hemisphere (to the right of the trade winds) and southwestward in the southern hemisphere (to the left of the trade winds), resulting in a divergence of water away from the equator. Surface water pulled away from the equator is replaced by rising water from the deeper ocean. This upwelling of cold water forms a cold band, in the middle of otherwise warmer water, that looks like a tongue extending from Peru and Ecuador out into the open ocean (shown in blue in Figure 9.18). Scientists commonly refer to this feature as the equatorial **cold tongue**.

The easterly trade winds also drag surface waters westward and induce a higher sea level, by about 40 cm, in the Western Pacific Ocean, as water accumulates in the western part of the ocean basin. This impacts the upper layer of warmer ocean water. This upper layer, a few hundred meters thick, is separated from the deeper cold ocean waters by a transition region called the **thermocline** (see Figure 9.19). In normal conditions, due to upwelling of deep, cold water in the eastern part of the basin, the thermocline slopes upward as it approaches South America (Figure 9.19(a)). In contrast, the thermocline sits lower in the western part of the basin, resulting in a deep pool of warm water in the Western Pacific Ocean called the **Pacific warm pool**. (Upwelling also occurs in the western part of the basin, but the mixed layer of warm water is much deeper, so that the cooling effect is small.)

The alert reader will have recognized a familiar setup. With warmer water in the Western Pacific and colder water in the Eastern Pacific due to upwelling, the conditions are ripe for a large-scale circulation driven by the temperature contrast (Figure 9.19(a)). Indeed, we observe lower pressures over Indonesia, along with convection, thunderstorms, and precipitation, and higher pressures over the cold tongue, along with subsidence and primarily low clouds.

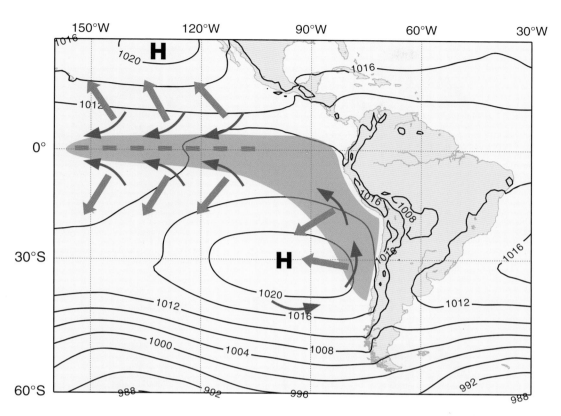

Figure 9.18. Formation of the cold tongue. The thin blue arrows represent the wind blowing northward along the coast of South America and converging along the equator. The thick gray arrows show the horizontal displacement of surface water, at a right angle to the wind. The blue cold tongue results from the upwelling of cold water from the deep ocean.

This large-scale east–west circulation takes place over thousands of miles and is referred to as the **Walker circulation**, after British scientist Sir Gilbert Walker (1868–1958), who first noticed the correlation between the Western and Eastern Pacific while analyzing pressure data to understand and predict the monsoon in India.

If the cold tongue is well known to local fishermen in Peru and Ecuador, it is also well known that temperatures and weather conditions change drastically every few years, following an irregular cycle of 2 to 7 years. Ocean temperatures rise, fish disappear, and heavy rains affect the coastal regions. In the meantime, Indonesia experiences drier conditions as convection moves eastward over the ocean. Because these episodes tend to start around Christmas, fishermen used to refer to them as **El Niño**, the child (i.e.,

the Christmas child, Jesus). We now know that these events are part of a large-scale oscillation involving both the atmosphere and the ocean, and we refer to it more generally as the "El Niño/Southern Oscillation" (**ENSO**).

ENSO involves a positive feedback loop between surface winds and upper-ocean conditions. The trade winds, for example, are observed to weaken during an El Niño event (Figure 9.19(b)). As a result, upwelling is reduced along the equator, which weakens the normal cooling process of surface water in the cold tongue. Furthermore, with weaker trade winds, the higher sea level in the Pacific warm pool is no longer sustained (a net force directed to the east), so that warm water begins to push eastward, further warming the water in the upper eastern ocean. As a result, the temperature contrast between Indonesia and the

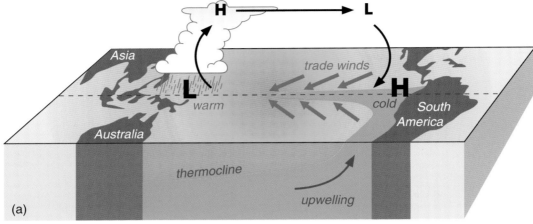

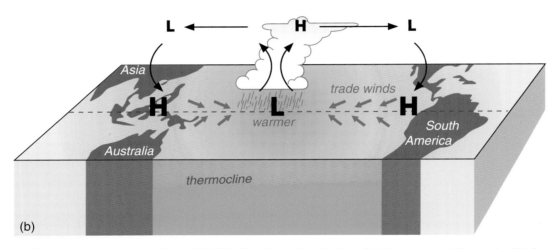

Figure 9.19. Schematic representation of El Niño/Southern Oscillation. (a) Normal conditions: the Walker circulation spans the entire Pacific basin. (b) El Niño: the cold tongue weakens or disappears, the Walker circulation weakens, and a secondary circulation forms over the western half of the Pacific basin.

Eastern Pacific, which sustains the Walker circulation, is also much reduced. In the process, low pressure and convection move eastward toward the Central Pacific, as Indonesia transitions to high pressure and subsidence, which explains the dry conditions experienced there during an El Niño event. The thermocline also levels out, as upwelling is reduced and warm water invades the Eastern Pacific (Figure 9.19(b)).

With a weaker Walker circulation, the trade winds, which really constitute the surface branch of the circulation, also weaken and we are back to where we started, with ever more weakening. This positive feedback loop, once initiated, shifts the whole system from normal to El Niño conditions, and can bring about drastic weather changes over large portions of Earth.

One might ask, when and why does the feedback loop stop? Or why does El Niño not keep growing indefinitely? To answer that question, we need to return to the ocean. As the trade winds blow over the Pacific Ocean, they drag the surface waters along and push them westward. To visualize how the water might pile up in the Western Pacific, one can think of a bathtub in which we are pushing the water to one side

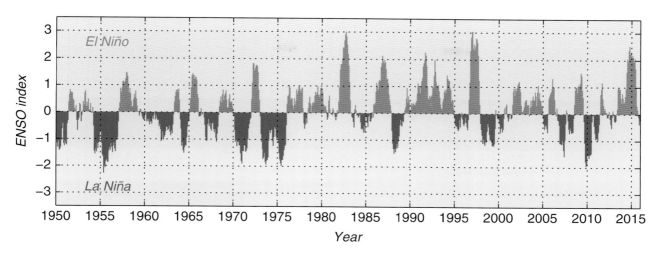

Figure 9.20. Variations of the most commonly used ENSO index since 1950. A positive index (in red) indicates El Niño conditions, while a negative index (in blue) indicates La Niña conditions.

with our hands. The water level will rise on that side. Similarly, the water level rises in the Western Pacific during normal conditions. This increase in sea level height is subtle (on the order of about 40 cm), but we can now measure it precisely with satellite instruments called altimeters. In the same way that our bathtub water will not remain piled up on one side of the tub if we stop pushing, the warm waters of the Western Pacific Ocean will start sloshing back toward South America as soon as the trade winds weaken and stop dragging the surface waters westward. The positive feedback will continue as the warm water propagates eastward on timescales of several months. But one can see that there is an inherent limit to the displacement of the warm water, due to the size and shape of the ocean basin and other dynamical constraints. El Niño corresponds to the maximum eastward extent of the warm water. When that natural limit is reached, heat from the warm water that has spread across the equatorial Pacific is radiated to space, the trade winds start dragging the warm surface waters westward again, and "recharge" the system, starting another cycle.

By analogy with El Niño, scientists have named the opposite phase of the oscillation, when the trade winds are unusually strong, the cold tongue is at its maximum extent, and warm surface waters are displaced to the west, **La Niña**. They have designed a number of

indices that indicate the extent to which the tropical Pacific ocean–atmosphere system is experiencing an El Niño or La Niña phase. These indices are typically based on measurements of atmospheric pressure, winds, and sea surface temperature. The most commonly used **ENSO index** combines multiple measurements and is shown in Figure 9.20. Positive values of the index indicate an El Niño phase, while negative values indicate a La Niña phase. A large positive value such as occurred in 1982–1983 and in 1997–1998 indicates a very strong, or *historic*, El Niño.

It is apparent in Figure 9.20 that ENSO follows an irregular cycle, with some phases being as short as 1 year, while others are as long as 6 years. A full cycle through El Niño and La Niña takes place over roughly 2 to 7 years, and is determined by complex ocean–atmosphere dynamics that involve more than just the tropics, and more than just the Pacific Ocean.

Because ENSO involves both atmosphere and ocean, it is called a coupled oscillation. Indeed, the displacement of water would not be possible without the trade winds, and the Walker circulation would not be possible without the ocean temperature contrast. Thus, both components are needed for the system to oscillate. Furthermore, we now know that ENSO affects weather patterns not only in the tropical Pacific, but also worldwide through what we call

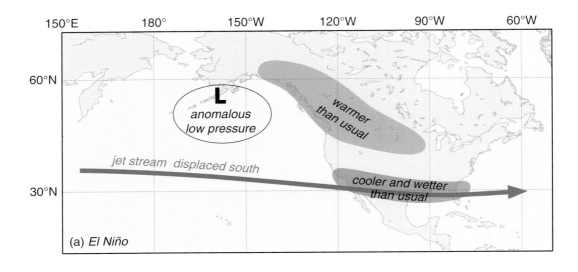

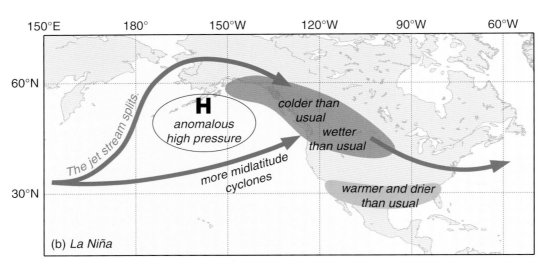

Figure 9.21. Impact of El Niño (a) and La Niña (b) on the weather of North America.

teleconnections – patterns of variation in the circulation of the atmosphere that are persistent over large distances. For example, during an El Niño event, tropical convection in the central equatorial Pacific Ocean distorts the jet streams in both hemispheres, producing large-scale waves on the jet stream. As a result, surface pressures are anomalously low over the Gulf of Alaska, which brings warmer temperatures than usual to the Pacific Northwest and Western Canada, while the jet stream is displaced to the south of its usual position, channeling storms toward California and Florida, where it is colder and rains more than usual (Figure 9.21(a)). By contrast, during a La Niña event, surface pressures are anomalously high over the Gulf of Alaska, which tends to split the jet stream into a northern branch, which brings cold, dry air to Western Canada, and a southern branch that steers midlatitude cyclones toward the Pacific Northwest, where it rains more than usual (Figure 9.21(b)). In the meantime, the Southern United States experience unusually dry weather.

El Niño also affects the frequency and intensity of tropical cyclones. Indeed, as we will see in Chapter 12, tropical cyclones are powered by the latent heat released in their core during condensation, the water vapor itself originating from evaporation of warm ocean water. This might explain why the warmer Pacific Ocean seems to fuel more intense tropical cyclones during El Niño events. However, tropical cyclones are also sensitive to the vertical wind shear (i.e., the change in wind speed with height – see Chapter 12), as strong shear disrupts their vertical structure and prevents self-organization. This is why the frequency of hurricanes decreases over the Atlantic Ocean during El Niño events, since the displacement of the jet stream to the south causes stronger upper-level winds and a stronger vertical wind shear over the latitude band where Atlantic hurricanes develop.

Thus, it appears that, through the interconnectedness of the general circulation, many large-scale weather patterns, and in particular large-scale oscillations, potentially affect each other through teleconnections. ENSO is by no means the only oscillation that animates the atmosphere. Other oscillations exist in the Atlantic Ocean (the North Atlantic Oscillation, NAO), in the North Pacific Ocean (the Pacific/North American pattern, PNA), and in the South Pacific Ocean (the Pacific/South American pattern, PSA), among others. Their mechanisms and interactions are not yet fully understood, and they remain active research topics in the atmospheric science community.

Summary

Time-averaged maps of weather variables (temperature, pressure, wind, precipitation, etc.) reveal semi-permanent features of the general circulation of the atmosphere, such as the region of strong temperature gradient found in each hemisphere between 30° and 60° latitude.

The polar front is associated with an increase in the speed of the westerlies from the surface up to the tropopause, where the westerly winds are maximum. This wind speed maximum is found in both northern and southern hemispheres, and is referred to as the **jet stream**. Two westerly jets often form in each hemisphere, a **polar jet** and a **subtropical jet**, but they sometimes merge to form a single jet stream, as is often the case over the North Atlantic Ocean and North Pacific Ocean.

The polar front and the jet stream are strongest in the winter hemisphere, due to the stronger temperature contrast between the tropics and the poles.

At the surface, average mean sea level pressure maps reveal semi-permanent low pressure regions in the midlatitudes, such as the Aleutian Low in the North Pacific Ocean and the Icelandic Low in the North Atlantic Ocean, and semi-permanent high pressure regions in the subtropics, such as the Pacific High and the Bermuda/Azores High in the northern hemisphere.

To account for these features, we discussed a "single-cell" model composed of an equator-to-pole Hadley cell in each hemisphere (i.e., an overturning vertical circulation spanning the depth of the troposphere). While this model explains the convergence of the trade winds, as well as cloudiness and precipitation at the **Intertropical Convergence Zone** (ITCZ), it does not account for the midlatitude westerly jets.

The "three-cell" model is composed of three overturning circulations in each hemisphere: a **Hadley cell** from the equator to 30° of latitude, a **polar cell** spanning 60–90° latitude and an intermediate region spanning 30–60° latitude, referred to as the **Ferrel cell**. The three-cell model explains the presence of subsidence and high pressure regions in the subtropics, where most of the major deserts are found.

The seasonal displacement of the ITCZ, combined with additional land–ocean temperature contrasts and dynamical effects, explain the alternating wet and dry seasons of the **monsoon** in India, South Asia, and sub-Saharan Africa.

In the equatorial Pacific Ocean, an interaction between the atmosphere and the ocean takes place that creates semi-periodic oscillations on timescales of 2 to 7 years, referred to as **El Niño/Southern Oscillation** (ENSO). **Upwelling** of deep, cold water off the coast of South America and along the equator creates a region of relatively colder ocean temperatures (the so-called **cold tongue**), which, in combination with the Pacific warm pool found in the western part of the basin, causes an east–west temperature contrast and a zonal, overturning circulation (the **Walker circulation**). Roughly every 2 to 7 years, a positive feedback mechanism weakens the Walker circulation, allowing warm water from the warm pool to slosh eastward, thus shrinking or replacing the cold tongue – a phenomenon known as **El Niño**. When the trade winds become stronger, the cold tongue intensifies, and warm water moves westward to the warm pool – a situation referred to as **La Niña**.

Although confined to the equatorial Pacific Ocean, ENSO influences weather patterns worldwide through **teleconnections**, as do other large-scale oscillations in other regions of the world.

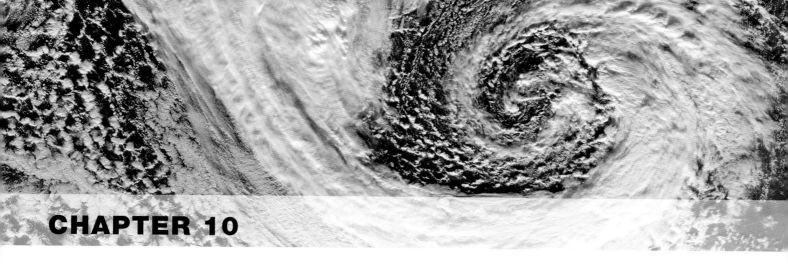

CHAPTER 10

Air Masses, Fronts, and Midlatitude Cyclones

Wind, clouds, rain... Most midlatitude weather is a result of the movement of warm tropical air poleward and cold polar air equatorward as the atmosphere acts to reduce the strong temperature gradients in the middle latitudes. The contrast of warm and cold air masses is most pronounced along warm and cold fronts, where most of the weather (clouds and precipitation) is found. We will now build a full picture of the midlatitude, or extratropical, cyclone, the weather system in which this air mass encounter is occurring.

We now have all the pieces of the midlatitude cyclone puzzle. As we get ready to assemble them and build a full story of cyclone development, let us review each component carefully.

10.1 Air Masses

When a large region of air spends several days over a land or ocean surface with relatively uniform properties, the air acquires the properties of the surface. In particular, it acquires the local temperature and moisture content of this *source region*. This large region of air of relatively uniform temperature and humidity is called an **air mass**. After some time spent over its source region, the air mass may be set in motion by

weather systems to transport its heat and moisture content to remote locations.

It is convenient to classify the main source regions into **polar** (with abbreviation **P**) vs. **tropical** (**T**), and **continental** (with prefix **c**) vs. **maritime** (**m**), depending on whether the air mass spent time at high or low latitude, and over land or ocean. Thus, by combination, four main air mass types can be observed, as illustrated in Figure 10.1 for North America:

▶ **Continental polar** (**cP**): cold, dry, stable air that originates over high latitude land areas, such as Canada or Siberia. When the air mass forms at very high latitude, we sometimes refer to it as **continental Arctic** (**cA**).

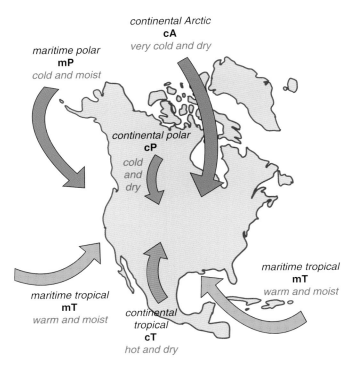

Figure 10.1. Main air mass types contributing to midlatitude weather in North America.

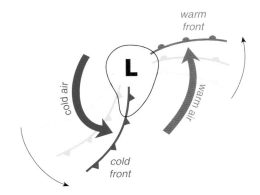

Figure 10.2. Cyclonic rotation of the cold and warm air masses, as well as associated fronts, around the low pressure center of a midlatitude cyclone.

▶ **Maritime polar** (**mP**): cool, moist, and usually unstable air that originates over high latitude oceans, such as the North Pacific or North Atlantic Oceans. Such air masses can have a continental polar origin, explaining their low temperatures, from where they move over the ocean and acquire water vapor by evaporation. Recall, however, that all is relative, as cold air contains only a small amount of water vapor and quickly reaches saturation.

▶ **Maritime tropical** (**mT**): warm, moist, and usually unstable air that originates over the low latitude oceans, such as the subtropical Pacific or Atlantic Oceans.

▶ **Continental tropical** (**cT**): warm, dry, stable air that originates over continental deserts.

Due to the heat contrast between the equator and the poles, and as a result of the natural tendency of the atmosphere to redistribute energy around the planet to reduce heat imbalance, tropical air masses

tend to move poleward overall, while polar air masses tend to move equatorward overall. (In other words, warm air moves toward colder locations and cold air moves toward warmer locations.) This exchange of air masses takes place in the midlatitudes in the form of a large weather system called a **midlatitude cyclone**, in which the two air masses spin around a central region of low pressure. The rotation is counterclockwise in the northern hemisphere and clockwise in the southern hemisphere, and is said to be *cyclonic*. Note that the rotation is consistent with geostrophic balance and the wind direction around the low pressure center (Figure 10.2).

10.2 Fronts

Where air masses are drawn into the circulation associated with the cyclone, there exist regions of sharp transition where temperature and humidity change abruptly. Temperature might increase from a low value characteristic of a polar air mass to a high value characteristic of a tropical air mass. Similarly, the dew point temperature might increase abruptly from a low value characteristic of a continental air mass to a high value characteristic of a maritime air mass. These regions of transition between air masses are called **fronts**, or frontal regions, and can be associated with intense weather activity within midlatitude cyclones. On a surface map, fronts are located on the warm edge of the region of maximum temperature gradient.

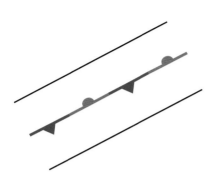

Figure 10.3. Representation of a stationary front on a surface map.

The "front" terminology came about in Europe in the early twentieth century, when the continent was at war and armies were fighting along a "front line". By analogy, the region where cold polar air was perceived to "advance against" warmer air was called a **cold front**, while the region where warm tropical air was perceived to "advance against" cooler air was called a **warm front** (Figure 10.2). On a weather map, cold fronts are represented with blue lines and blue triangles pointing in the direction of motion of the cold air, while warm fronts are represented with red lines and red semi-circles pointing in the direction of motion of the warm air.

10.2.1 Stationary Fronts

Fronts develop where two air masses come in contact. This can happen in the midlatitudes where warm tropical air lies next to cold polar air, with small movement in the air masses. We refer to this contrast as a **stationary front** and represent it on a weather map with alternating blue triangles and red semi-circles facing in opposite directions (Figure 10.3). If the two air masses are set in motion and start rotating cyclonically around an incipient low pressure center, the stationary front evolves into a different structure. It is then useful to distinguish three types of fronts.

10.2.2 Cold Fronts

A cold front is formed where cold air moves into warmer air (Figure 10.4(a)). Cold fronts are char-

acterized by a sharp contrast in temperature and moisture. The wind direction shifts abruptly from the cold to the warm air, which corresponds to a pressure trough (Figure 10.4(b)). This sharp turn in the isobars provides good guidance as to the position of the cold front on a surface map. Cold fronts are usually narrow and coincide with clouds and precipitation, typically of the convective type. Note how the cloud band is concentrated along the cold front in Figure 10.4(a). Clouds form as warm air is forced upward ahead of the cold front and tend to produce a band of showers ahead of the front (Figure 10.4(c)). In some cases, clouds can also be observed behind the surface cold front on a satellite image, in which case they can produce precipitation behind the cold front.

10.2.3 Warm Fronts

A warm front is formed where warm air moves into cooler air (Figure 10.5(a)). The pressure trough and wind shift are usually not as pronounced at the warm front as they were at the cold front, but the pressure trough still provides a good indication as to the location of the warm front.

Warm fronts also tend to move more slowly than cold fronts. There are several factors that control the speed of movement of fronts, including the evolution of the wind field, which may squeeze or pull apart isotherms in a way that affects the speed of the front. Ignoring that process, which is more advanced than we can discuss here, we may regard the front simply as a moving air mass boundary. In that case, at the warm front, warm air is ascending over the cold air mass (overrunning), which means that, at the surface, the front moves from erosion of cold air from above; this retreat of cold air is a slow process. In contrast, the cold air behind the cold front is advancing, and not retreating. It wedges underneath the warm air, easily displacing the warm air due to the density difference between these two air masses. The sector delimited by the two fronts, called the **warm sector** (Figure 10.5(b)), often shrinks with time as the air masses "wrap up" around the low center, and as a result we say that the warm sector *closes*.

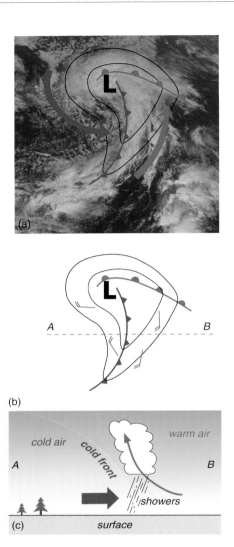

(a)

(b)

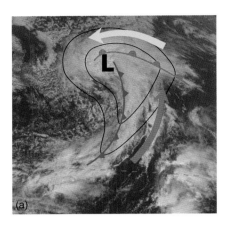

(a)

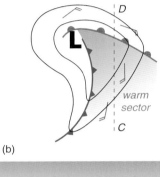

warm sector

(b)

Figure 10.4. Anatomy of a cold front in a midlatitude cyclone. (a) Cyclonic rotation of cold and warm air around a low pressure center, superimposed on satellite imagery. (b) Pressure trough and wind shift corresponding to the cold front. (c) Vertical cross-section through the cold front along line *AB* in (b).

Figure 10.5. Anatomy of a warm front in a midlatitude cyclone. (a) Cyclonic rotation of warm and cooler air around the low pressure center, superimposed on satellite imagery. (b) Warm sector and wind shift corresponding to the warm front. (c) Vertical cross-section through the warm front along line *CD* in (b).

As the warm air rises gently over the cold air mass, it creates a more extensive area of clouds and precipitation with a milder slope than cloud structures encountered along a cold front. Stratiform clouds tend to form, producing lighter and longer-lasting precipitation (Figure 10.5(c)). They form an extensive cloud deck, which is easily discernible on satellite images, ahead of the warm front. The combination of an extensive cloud deck with a narrow tail of convective clouds along the cold front gives the cyclone the appearance of a comma in satellite imagery.

10.2.4 Occluded Fronts

After an incipient low has formed (Figure 10.6(a)), the midlatitude cyclone develops (Figure 10.6(b)) and deepens (Figure 10.6(c)), provided the appropriate upper-level conditions are present (see Section 10.3.2). As the cold and warm air wrap around the low pressure center, the warm sector narrows and becomes separated from the low pressure center, now isolated in the cold air (Figure 10.6(d)). At this mature stage, the warm air near the cyclone has been

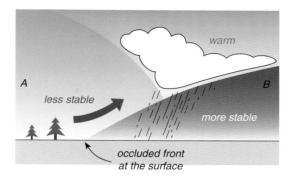

Figure 10.7. Frontal structure at the occluded stage. (The cross-section refers to line *AB* in Figure 10.6(d).)

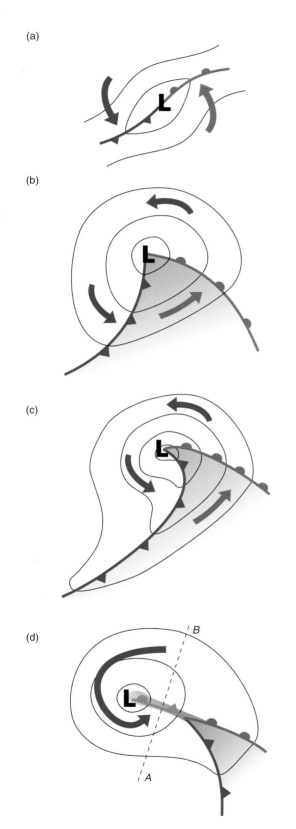

Figure 10.6. Four stages of the life cycle of a midlatitude cyclone: (a) incipient, (b) deepening, (c) mature, and (d) occluded.

pushed up above the surface. This is better seen on a cross-section, such as that shown in Figure 10.7. The advancing cold air, being typically less stable than the warmer air ahead, slides *over* it, producing a frontal surface slanting forward from the occluded front at the surface. It is not uncommon for the low pressure center to continue deepening after occlusion has started and to produce extensive clouds and a wide range of precipitation types.

The configuration depicted in Figure 10.7 is also called a *warm occlusion*, because the overruning is reminiscent of the overrunning of warm air over cold air observed along a warm front. In theory, a *cold occlusion* may also occur, in which the advancing cold air is more stable and the occluded frontal surface slopes rearward with height; however, cold occlusions are rare and very few have been documented.

10.2.5 Large-Scale Influences on Cyclone Structure, and the T-bone Model

The previous description of air masses and fronts is based on an early model of cyclone development by Norwegian meteorologists at the beginning of the twentieth century (see Appendix 10.2). While providing a generally good and useful description of midlatitude weather, this model is not universal and does not account for factors that affect the structure of cyclones and fronts. For example, the structure of the jet stream has a strong influence on the life cycle of cyclones. Changes in the jet stream winds in the east–west direction can elongate or shrink the

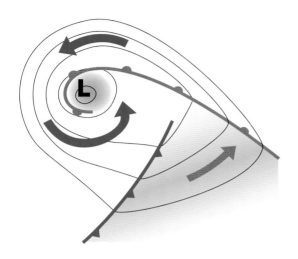

Figure 10.8. "T-bone" model: an alternative to the Norwegian model of cyclone and frontal development.

fronts in that direction, and this differs systematically for cyclones developing at the eastern and western ends of ocean basins. For example, at the western end of the basin, where the jet stream winds increase toward the east, warm fronts tend to be more extensive than at the eastern end, where the jet stream winds tend to decrease toward the east. Changes in jet stream winds in the north–south direction have been found to control how the fronts wrap up around the cyclone, with cyclones on the north side of the jet stream getting more intense "wrapping" than those to the south. There is also evidence that some cyclones, especially over the ocean, lack a clear occlusion as described above, and atmospheric scientists developed alternative models. One such alternative is the Shapiro–Keyser model, named after Dr. Mel Shapiro and Professor Daniel Keyser, and sometimes also referred to as the *T-bone model*. In this model, the cold front tends to be at a right angle to the warm front, producing a structure reminiscent of a T (Figure 10.8). This model better depicts situations for strong oceanic cyclones, in which the air masses are wrapped completely around the low pressure center, isolating a core of warm air in the middle of colder air ("warm core seclusion"), while the warm front bends around the low pressure center ("bent-back front"). *Sting jets* (especially damaging winds descending from the midtroposphere) are sometimes associated with

the end of the bent-back front, while cloud bands can be observed to spiral into the low pressure region on satellite images.

10.3 Midlatitude Cyclone Development

How do all these elements fit in the same picture? How do midlatitude cyclones come about? To answer these questions, we need to understand the full development of a cyclone, from the initial disturbance to decay, as a three-dimensional weather system spanning the entire depth of the troposphere.

10.3.1 The Life Cycle of a Midlatitude Cyclone

We first need to return to the notion of heat imbalance. With a constant gain of heat near the equator and loss of heat near the poles, the atmosphere would not be a balanced system in the absence of a mechanism to redistribute heat from the source region to the sink region. The natural tendency of any system out of balance is to seek to return to equilibrium, and, in the case of the atmosphere, midlatitude cyclones are a primary mechanism that maintains equilibrium by moving heat poleward. Since the heat imbalance is constantly recreated by the uneven absorption of solar radiation around Earth, there is a persistent driving force creating midlatitude cyclones.

As we discussed in Chapter 9 (Sections 9.2 and 9.3), due to Earth's rotation, the poleward redistribution of heat is not accomplished by a single Hadley cell, but rather takes place in midlatitude cyclones and the exchange of air masses that occurs with these features. Because of Earth's rotation, the westerly jet streams are intrinsically unstable: they become wavy, and eventually break down into individual cyclones. Each cyclone goes through a series of phases generally described as the *life cycle* of midlatitude cyclones.

Most often, these cyclones originate from disturbances in the westerly jet streams, troughs of low pressure (low heights), called short-wave troughs (or "short waves"), that cause the pressure to fall at the

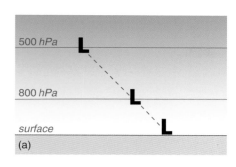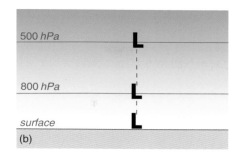

Figure 10.9. Deepening lows tilt westward with height (a), whereas occluded lows are vertically aligned (b).

surface, as we will discuss later in this chapter. Often there is no pre-existing surface front when this happens, and fronts develop along with the cyclone. In some cases, however, the life cycle begins with a wave cyclone on a pre-existing surface front, which is the sequence described by the **Norwegian cyclone model**. For the sake of simplicity, we proceed with a description of the Norwegian model, but note that the case without an initial front is described by stages (2) through (5) after the formation of a surface low associated with an upper-level disturbance:

(1) **Wave cyclone**: A weak pressure disturbance appears as warm air starts to move poleward and upward, while cold air starts moving equatorward and downward (Figure 10.6(a)). Convergence and upward motion cause a patch of clouds to form poleward of the developing warm front, which is often the first hint of the developing cyclone on satellite imagery.

(2) **Deepening cyclone**: The central pressure falls and the low is now delimited by several closed isobars (Figure 10.6(b)). The cold and warm fronts are far apart and the warm sector is broad.

(3) **Mature stage**: The cyclone is fully developed with well-defined cold and warm fronts and a relatively low central pressure (Figure 10.6(c)). It has an easily recognizable comma shape on satellite imagery.

(4) **Occluded stage**: Cold air wraps around the low pressure center, which becomes isolated in the cold air (Figure 10.6d). The warm sector is increasingly separated from the low pressure center, but a vestigial axis of relatively warm air remains along the occluded front.

(5) **Decaying stage**: The cyclone decays as the air column fills in and the central pressure increases. Clouds persist but gradually dissipate as they wrap around the low center.

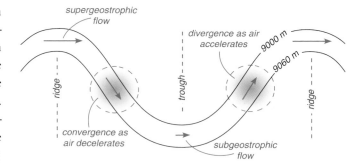

Figure 10.10. Areas of convergence and divergence within a 300 hPa trough.

This life cycle can take from about 2 to 6 days, depending on the time of year, the strength of the midlatitude temperature gradient, and environmental conditions. The cyclone, being embedded in a mean midlatitude flow that is westerly overall, moves eastward while going through the life cycle. In the southern hemisphere, midlatitude cyclones also move eastward, but their structure is a mirror image of their northern hemisphere counterparts with respect to the equator (see Appendix 10.1).

This model of cyclone development was first proposed by meteorologists in Bergen, Norway, at the beginning of the twentieth century, and is usually referred to as the *Norwegian model* from the *Bergen school*, or the *Bjerknes model*, after Jacob Bjerknes (1897–1975), who wrote a seminal article about polar front theory in 1922. As general as it is, this model gives us some level of predictability as to how a pressure disturbance will evolve over the days following initial development. It does not, however, tell us *where* a pressure disturbance will appear and which of these pressure disturbances will evolve into fully developed

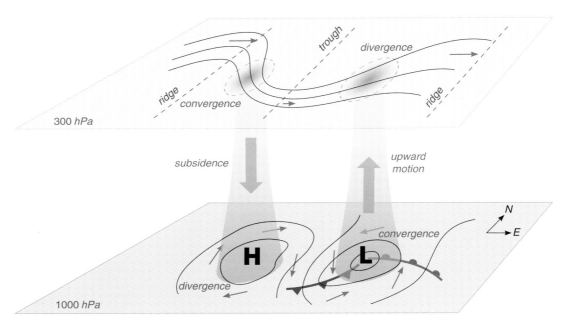

Figure 10.11. Superposition of upper-level convergence and divergence patterns over a surface anticyclone and a deepening cyclone.

cyclones. To answer these questions, we need to make some additional observations and return to the upper levels.

10.3.2 Vertical Structure of Cyclones

A comparison of the pressure distribution at various levels of the atmosphere is very instructive. It reveals that the pressure minimum characteristic of a deepening cyclone is not found at the same location on different pressure levels. The upper-level minimum is found westward of the surface minimum during development. In other words, if we picture the cyclone as a spinning vortex with a core of low pressure, the core of a deepening cyclone is not vertical, but tilts westward with height (Figure 10.9). By contrast, the core of a cyclone that has reached the decaying stage is vertically aligned, or tilted eastward with height. This suggests that we need to investigate the upper-level flow, and in particular the alignment of upper-level troughs with surface lows, to understand the timing and location of cyclone development.

Figure 10.10 shows a schematic representation of a 300 hPa trough flanked by two ridges. You might recall from Chapter 8 that the flow is supergeostrophic around a ridge and subgeostrophic around a trough, as shown in Figure 10.10 by arrows of different sizes. If we now follow the air flow along a "channel" delimited by two isobars, we note that the air enters the channel with greater speed, and decelerates as it reaches the trough. This causes a **convergence** of air between the ridge and the trough (i.e., there is more air entering the channel than leaving it). Conversely, further downstream, east of the trough, the air accelerates as it flows into the next ridge. This causes a **divergence** of air (i.e., there is more air leaving the channel than entering it). This dual pattern of convergence–divergence has two major effects:

(1) Net convergence adds air to the air column, which means that the pressure underneath rises. Conversely, net divergence removes air from the air column, which means that the pressure underneath falls.

(2) For reasons similar to those invoked in earlier chapters, air cannot accumulate indefinitely where there is convergence and must take a new direction. Since we are at 300 hPa, at the tropopause, with a strong temperature inversion above and a very stable stratosphere preventing upward motion, the converging air cannot rise, and so it sinks downward. Therefore, the region of convergence is a region of *subsidence* as well as rising surface pressure. Conversely, diverging air is

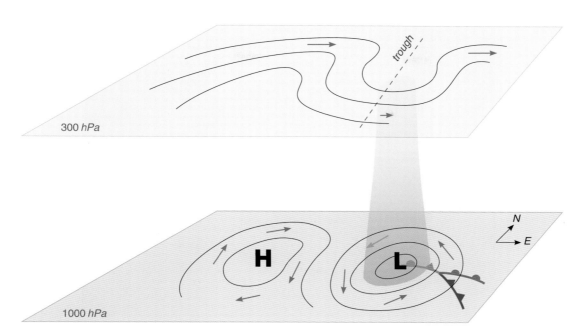

Figure 10.12. Displacement of the upper-level trough over the surface low as the midlatitude cyclone becomes occluded and decays.

replaced by air rising from below, and the region of divergence is a region of *upward motion* as well as falling surface pressure.

There we have a good hint as to why low pressure tilts westward with height for developing cyclones, and why the precursors for development are troughs in the jet stream. If the surface cyclone is to deepen, it needs to be under the region of upper-level divergence, east of the upper-level trough. And that is indeed what we observe, as shown schematically in Figure 10.11. As the upper-level trough, moving from west to east, approaches the surface low from the west, the region of upper-level divergence is ideally positioned to cause upward motion (producing clouds) and decrease the central pressure of the surface low. In other words, it causes the surface low to deepen.

Surface winds are consistent with this picture. Recall that, due to friction, surface winds do not blow parallel to the isobars, as suggested by geostrophic balance, but flow toward lower pressure *into* the low, which results in surface convergence. Thus, in concert with the upper-level flow, surface winds tend to converge and to reinforce upward motion.

West of the upper-level trough, the inverse situation can be observed. Upper-level convergence

leads to subsidence, which coincides with the surface divergence typically observed in anticyclones, as surface winds tend to flow *away* from the high pressure (toward lower pressure).

As the cyclone deepens and matures, the upper-level trough moves *over* the surface low (Figure 10.12) and the region of upper-level divergence, which has moved further east, can therefore no longer sustain upward motion and decreasing pressure in the cyclone. Intensification stops, and the cyclone begins to decay when the upper-level trough continues eastward and the tilt reverses from what it was during cyclone development. Since the converging air at the surface no longer has an upper-level outlet, convergence adds air mass to the air column and the surface pressure increases. We say that "the low fills in". Friction gradually spins down the surface winds.

Note that upper-level divergence is necessary for a midlatitude cyclone to deepen and intensify, but it is not sufficient. If the atmosphere is too stable, it resists upward motion, even in the presence of upper-level divergence. Therefore, an additional condition for cyclone development is a relatively weak atmospheric stability in the region where the incipient low is forming. This helps explain why we find cyclones

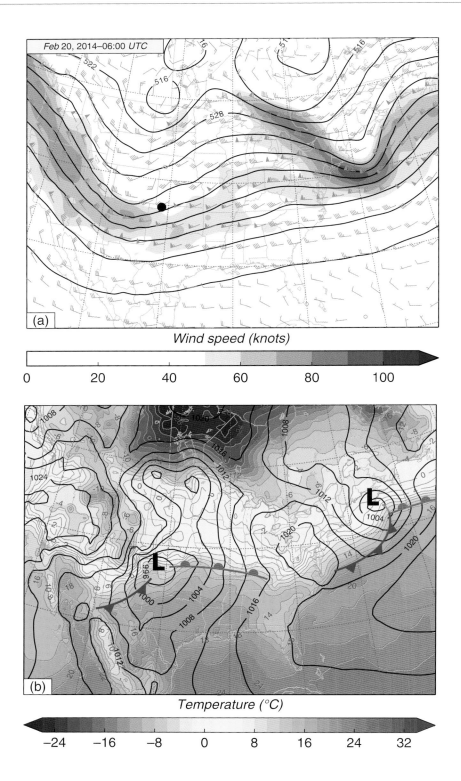

Figure 10.13. Incipient wave on February 20, 2014, at 06:00 UTC. (Top left) 500 hPa heights with winds and wind speed. The position of the surface low is indicated by a black dot. (Bottom left) Surface pressure and temperature. (Top right) Infrared satellite image with surface isobars and the position of the surface low, with cold and warm fronts. (Bottom right) Surface pressure, surface winds, and precipitation. (Note that there is more rain and snow in Canada than suggested in the figure because there are fewer weather stations in Canada than in the United States. See Figure 1.21.)

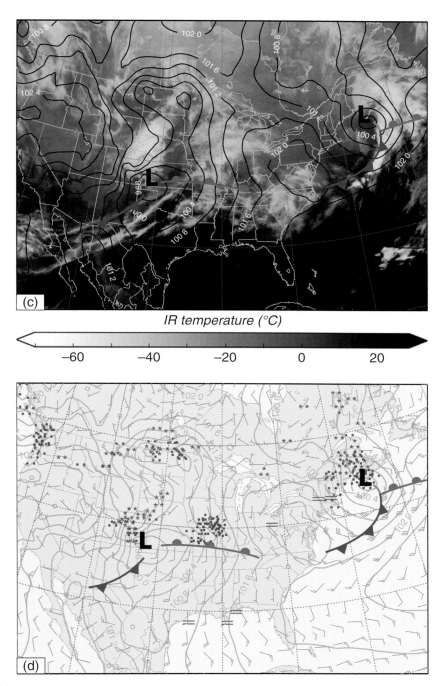

IR temperature (°C)

Figure 10.13. (cont.).

developing preferentially over the oceans, where during winter the relatively warmer water heats the air near the surface, which produces weak stability (see Section 3.4).

In the end, from wave cyclone to decay, the cyclone has gone through a full life cycle and has effectively exchanged warm and cold air across the midlatitude temperature gradient, which results in a net flow of heat toward the poles.

10.3.3 The February 2014 Cyclone

We can now verify how all these theoretical concepts apply to a real case: the February 2014 cyclone we have been studying throughout the book.

On February 20, 2014, at 06:00 UTC, a surface low pressure center has formed over the Texas panhandle. A small cyclonic circulation has started to develop, with small cold and warm fronts (Figure 10.13).

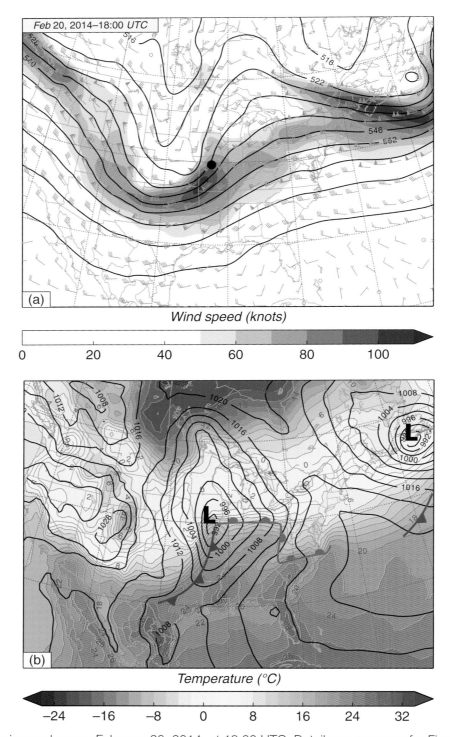

Wind speed (knots)

0   20   40   60   80   100

Temperature (°C)

−24   −16   −8   0   8   16   24   32

Figure 10.14. Deepening cyclone on February 20, 2014, at 18:00 UTC. Details are same as for Figure 10.13.

We have a nice example of an *incipient wave*. Another area of low pressure is found over North Dakota, with one center of low pressure along the border with Minnesota, in the location favored for development to the east of a weak trough at 500 hPa. Some snow is found north of our incipient wave along with the weak system over North Dakota. The infrared satellite image suggests little cloudiness along the cold front at this stage, and some stratiform clouds north of the warm front, corresponding to the rain showing in the surface weather station measurements (Figure 10.13(d)). The surface low is ideally located

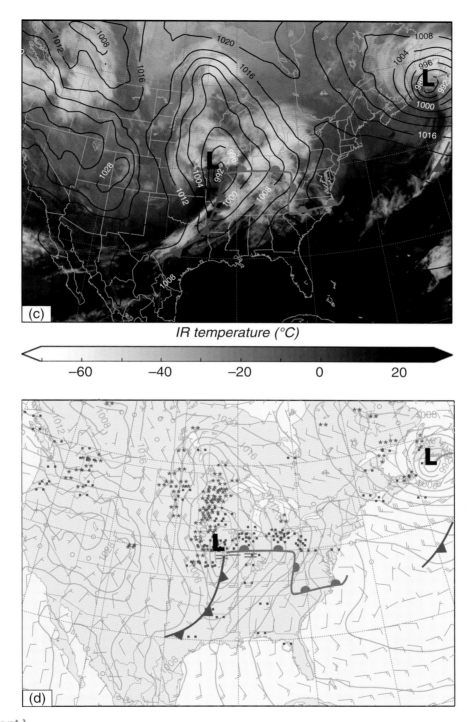

Figure 10.14. (cont.).

underneath an upper-level trough, east of the trough axis, where we expect upper-level divergence to enhance upward motion and deepen the surface low, as we just discussed.

Twelve hours later (Figure 10.14), the main cyclone has indeed deepened significantly. We can see more closed isobars, and stronger winds (i.e., tighter isobars). The fronts are better established, with a clearly defined warm sector, and a sharp turn of the wind at both the cold and the warm fronts (i.e., a sharper turn, or kink, in the isobars). Clouds are more extensive, with rain reported south of the Great Lakes,

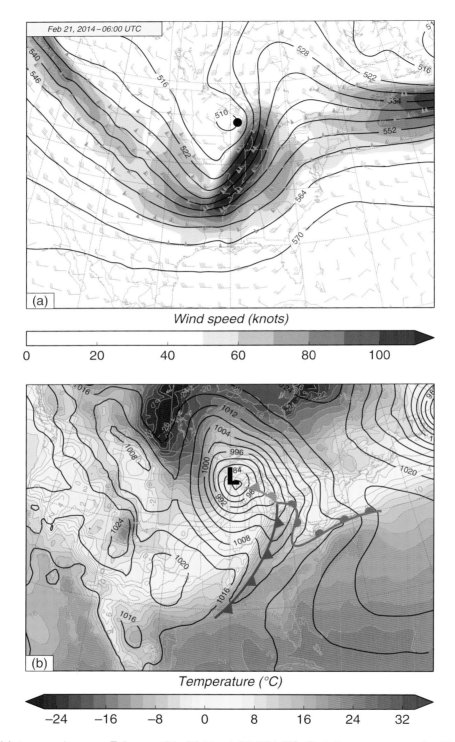

Figure 10.15. Mature cyclone on February 21, 2014, at 06:00 UTC. Details are same as for Figure 10.13.

and an extensive area of snow over Minnesota, Iowa, and Missouri, where cool air from the east ascends over very cold air from Canada. Snow is observed at the surface even though temperatures are above freezing in many locations. Recall that snowflakes can fall through about 300 m of above-freezing air before completely melting. The surface low is still ideally located underneath and east of the upper-level

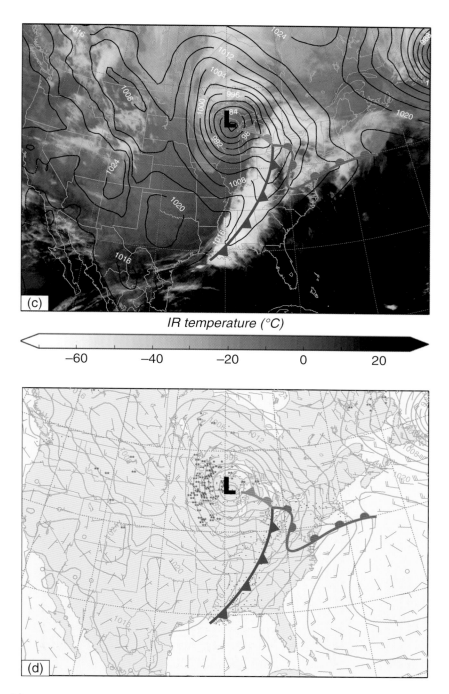

IR temperature (°C)

−60	−40	−20	0	20

Figure 10.15. (cont.).

trough, which will lead to further deepening. The second low to the north has become subsumed into the main cyclone, but still appears in the surface isobars as a trough extending north from the main cyclone into Minnesota.

On February 21, 2014, at 06:00 UTC, the cyclone has reached the mature phase (Figure 10.15). The central pressure has dropped below 980 hPa, and an occluded front has started forming where the cyclone is separating from the warm sector. Notice the axis of relatively warm air that extends along the occluded front from the low center to the "**triple point**," which is defined by the point of intersection of the warm, cold, and occluded fronts. A thick band of clouds is

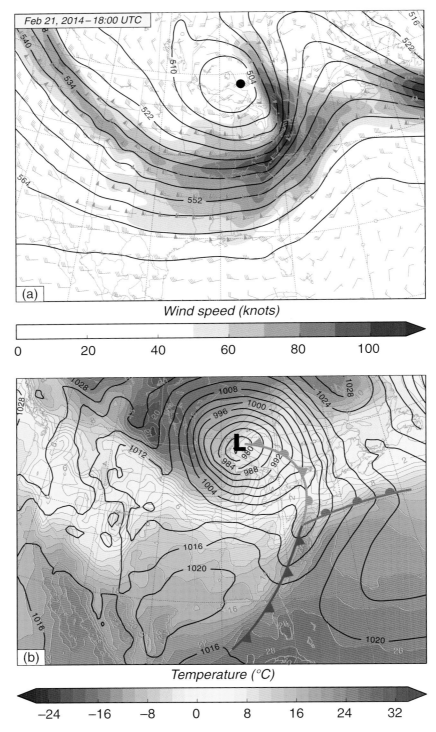

Figure 10.16. Decaying cyclone on February 21, 2014, at 18:00 UTC. Details are same as for Figure 10.13.

aligned with the cold front, producing rain ahead of the front, consistent with our cold front picture, in which the cold air lifts the warmer air ahead and causes the formation of clouds and precipitation (Figure 10.4). Rain continues ahead of the warm front, where air from the warm sector is overrunning colder air to the north, as well as snow west and north of the Great Lakes. The surface low is nearly underneath the

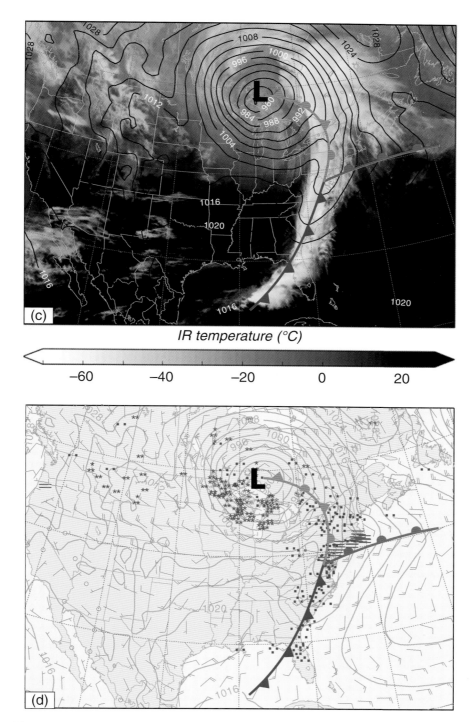

IR temperature (°C)

Figure 10.16. (cont.).

upper-level trough axis. Since it is leaving the area of upper-level divergence, we anticipate the end of the developing phase of the cyclone.

Twelve hours later (Figure 10.16) we find that a closed low has formed at 500 hPa, and that this upper-level low and the surface low are now stacked on top of each other, as described in Figures 10.9(b) and 10.12.

The central pressure of the surface cyclone is still below 980 hPa, which is similar to what it was 12 hours previously. Surface winds now circle completely around the low, along with cold air and snow over the Great Lakes. The cyclone is now fully occluded, with an axis of warm air connecting the low center to the warm sector at the triple point, and, from then on, the

(a)

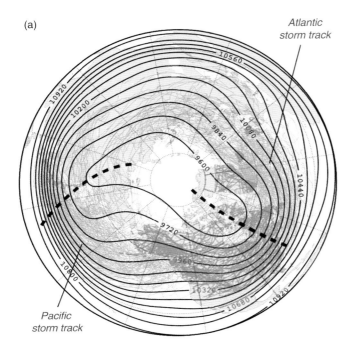

(b)

Figure 10.17. 500 hPa heights overlaid on 7 years of midlatitude cyclone trajectories (in blue). (a) In the northern hemisphere, storms tend to form southeast of the semi-permanent upper-level troughs (dashed lines) and to follow the Pacific and Atlantic storm tracks. (b) In the southern hemisphere, storms form all around Antarctica.

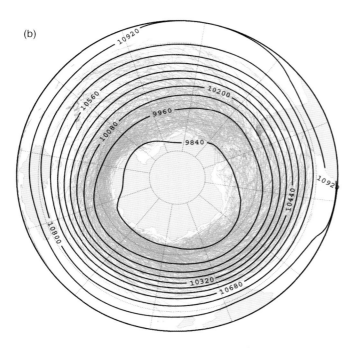

cold front will continue its eastward course detached from the main low.

This cyclone was one of the strongest cyclones to affect the area during the winter of 2014. It produced everything from heavy rain, to sleet, to freezing rain, and, of course, snow. The strong pressure gradients produced very strong winds, with up to 65 mph wind gusts. It spawned several thunderstorms ahead of the cold front, and even a couple of tornadoes were reported in Illinois.

10.3.4 Where do Cyclones Form?

We have established that midlatitude cyclones form on the midlatitude temperature gradient, in a region of relatively weak static stability downwind of a trough in the westerly jet stream. There are certain regions, however, where cyclogenesis (i.e., the genesis of cyclones) is enhanced: regions where the combination of factors leading to cyclone intensification is more optimal than others. You might recall, for example, the presence of semi-permanent upper-level troughs east of Asia and east of North America in our mean upper-level pressure maps (Figure 9.1). Even though the upper-level flow changes and evolves from one day to the next, and there is no stationary trough on a day-to-day basis, the fact that there exists a trough *on average* reveals that the occurrence of upper-level troughs is more pronounced in those two regions. It is largely due to the presence of major mountain ranges (Tibetan Plateau and Rocky Mountains) and land–ocean temperature contrasts associated with the Eurasian and North American continents.

Since ocean waters are relatively warm in these regions, the atmosphere is also relatively warm and moist, and therefore only weakly stable. This combination of factors makes these two regions prone to midlatitude cyclone development. Cyclones tend to form over the Western Pacific and West Atlantic and cross the ocean basin as they deepen and mature. We refer to these two regions of enhanced cyclonic activity as the **storm tracks** (Figure 10.17(a)).

In the southern hemisphere, land masses are not as prominent, and the Southern Ocean offers a continuous channel around Antarctica. Therefore, cyclogenesis is more uniformly distributed, with a relatively higher incidence of cyclones east of South Africa, southeast Australia, and the southern tip of South America (Figure 10.17(b)).

Cyclogenesis is also enhanced in the lee of mountain barriers such as the Rockies in North America and the Andes in South America, where prevailing westerly winds produce downslope warming and a trough of low pressure at the surface. In fact, our February 2014 cyclone was one such example of a low that formed in the lee of the Rockies.

Similarly, the low pressure trough corresponding to the cold front of a mature or decaying cyclone can serve as the environment for the birth of a new cyclone. The pressure is already locally lower and the trough has the characteristics of a front: if an upper-level trough comes along and produces divergence above the trough, it can produce an incipient wave, called a frontal wave. Given the right conditions, this incipient wave can grow into a full-fledged cyclone. We often see this behavior at the eastern end of the storm tracks, where mature cyclones leave long trailing cold fronts, which provide a fertile environment for upper-level troughs to begin the process anew.

In the end, knowing these areas of enhanced cyclogenesis allows us to recognize familiar patterns and to forecast better the formation and development of midlatitude cyclones.

Summary

Midlatitude cyclones form on temperature contrasts near the jet stream. Winds circulating around the cyclone distort and concentrate the temperature contrasts into fronts. The frontal region where cold air moves into warmer air is called a **cold front** and is characterized by a sharp temperature and moisture contrast along a pressure trough, a shift in wind direction, and usually a band of clouds and precipitation, often of the convective type.

The frontal region where warm air moves into cooler air is called a **warm front** and is also characterized by a pressure trough and a shift in wind direction, although usually not as pronounced as for the cold front. As the warm air rises gently over the cold air mass (**overrunning**), stratiform clouds tend to form, producing lighter and longer-lasting precipitation.

As the fronts become stretched and elongated, the low center separates from the warm sector, creating an **occluded front**. An axis of relatively warm air is found along the occluded front, connecting the low center to the triple point where all three fronts meet. Aloft, air that was behind the cold front has moved over the warm front due to the fact that the air below the warm front is relatively more stable (and therefore the air behind the cold front is easier to lift).

The midlatitude cyclone goes through a typical **life cycle** by which:

▶ a first **wave cyclone**, or incipient low, forms;

▶ the low **deepens** (i.e., the central pressure falls), and the cyclone intensifies (stronger winds, more clouds, and precipitation);

▶ it reaches a **mature stage** where it is fully developed and the rate of deepening often begins to slow;

▶ the low separates from the warm sector, the low pressure center migrates back into the cold air under the upper-level trough, and an **occluded front** forms; and

▶ the cyclone eventually decays (i.e., the central pressure increases).

When a surface low pressure center is located east of an upper-level trough axis, upper-level divergence makes the surface pressure fall and the cyclone intensify. Conversely, upper-level convergence causes subsidence and the development of high pressure at the surface, to the west of the upper-level trough axis. When the upper-level trough moves *over* the surface low, intensification stops and the cyclone may begin to decay.

Appendix 10.1   Southern Hemisphere Midlatitude Cyclones

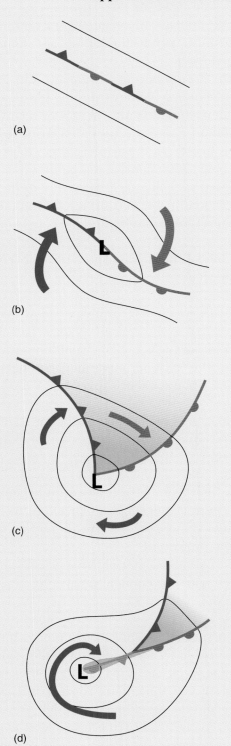

Recall that, in the southern hemisphere, the Coriolis force deflects air parcels to the left, our rules of geostrophic motion are reversed, and cyclones rotate clockwise. Therefore, we can flip all the schematics in Figure 10.6 upside down: they are really a mirror image of the northern hemisphere case with respect to the equator (see Figure 10.18). However, remember that the southern hemisphere jet stream is also composed of westerly winds (Figure 10.19) – do not flip the direction of the jet stream! Moreover, the upper-level waves that make the jet stream undulate also propagate eastward in the westerly flow; therefore, midlatitude cyclones overall progress from west to east, as in the northern hemisphere.

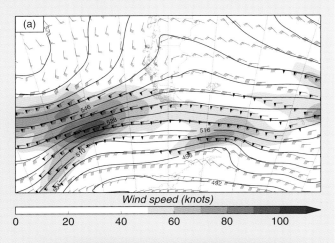

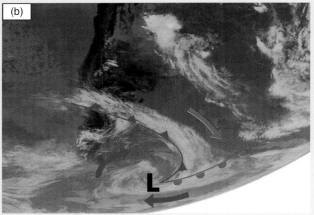

Figure 10.18.   Life cycle of a southern hemisphere midlatitude cyclone: incipient (a), deepening (b), mature (c), and occluded (d) stages.

Figure 10.19.   Example of a southern hemisphere midlatitude cyclone on February 20, 2014, at 18:00 UTC. (a) 500 hPa heights and winds. (b) Infrared image with surface fronts and general air mass circulation.

Appendix 10.2 The Bergen School of Meteorology

The Norwegian model of cyclone development was developed at the beginning of the twentieth century by a group of meteorologists working under the supervision of **Vilhelm Bjerknes** (Figure 10.20) and later his son, **Jacob Bjerknes** (Figure 10.21), who wrote the seminal paper "On the structure of moving cyclones" in 1919. Father and son laid the foundation for the Bergen school of meteorology, a group of meteorologists based at the University of Bergen, Norway, and developed a school of thought that has inspired much of modern meteorology.

Several of the Bergen school scientists have made significant contributions to the field of meteorology. **Tor Bergeron** (Figure 10.22) proposed the mechanism explaining the growth of ice crystals at the expense of liquid water droplets in cold clouds, known as the Bergeron process (see Chapter 7). **Sverre Petterssen** (1898–1974) is remembered for his D-Day forecast during the Second World War, and encouraging General Dwight Eisenhower to postpone the landing in Normandy from June 5 to June 6, 1944. **Carl-Gustaf Rossby** (Figure 10.23) founded the Chicago school of meteorology and advanced our understanding of large-scale atmospheric motions. In particular, he identified and characterized the undulations of the jet stream, now referred to as **Rossby waves**.

Figure 10.20.
Vilhelm Bjerknes
(1862–1951).

Figure 10.21.
Jacob Bjerknes
(1897–1975).

Figure 10.22.
Tor Bergeron
(1891–1977).

Figure 10.23.
Carl-Gustaf Rossby
(1898–1957).

CHAPTER 11

Thunderstorms and Tornadoes

Upwardly buoyant air parcels can produce powerful thunderstorms towering cumulonimbus clouds producing rain, hail, lightning, and thunder. When these thunderstorms rotate, they can also spawn tornadoes and cause some of the most severe weather on Earth. We will now describe the different stages of thunderstorm development, the formation of lightning and thunder, and will explain the circumstances and mechanism by which tornadoes can form at the base of a supercell thunderstorm.

Mild convection often produces innocuous fair-weather cumulus clouds. We recall that air parcels, if they are warmer than the environment, experience an upward buoyancy force that can lift them above the lifting condensation level (LCL), where their temperature drops below the dew point temperature, causing condensation and the formation of cloud droplets. When atmosphere is quite unstable, however, continued convection can produce tall cumuliform clouds along with hail and rain in the form of heavy showers. When the wind shear (i.e., the change in wind speed with height) is weak, what are referred to as *ordinary* storms develop. Such storms generally are not accompanied by severe

weather, but exceptions can occur in environments that are extremely unstable. When significant wind shear is present, highly organized storms can develop that can take the form of *supercells* or *squall lines* and produce more extensive severe weather. We will now describe each type of thunderstorm in detail.

11.1 Ordinary Thunderstorm

The development of a thunderstorm is usually preceded by the formation of one or several cumulus clouds that grow and dissipate, moistening the air

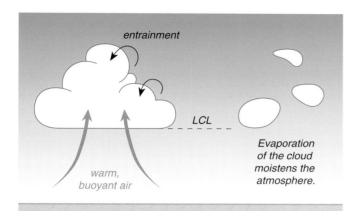

Figure 11.1. Moistening of the atmosphere by formation and evaporation of cumulus clouds.

in the process, and prime the atmosphere for subsequent storm development (Figure 11.1). The convective process being quite turbulent, the air parcels delimiting the cloud and containing a high level of moisture and cloud droplets are constantly mixing with drier air parcels outside the cloud. This mixing of cloud parcels with dry air is called **entrainment**. The entrainment of dry air into the cloud lowers the dew point temperature and causes some of the cloud droplets to evaporate. Thus, in the end, moisture has been transported upward by convection, from the surface to a higher region of the atmosphere, where it has increased the relative humidity. The first cumulus cloud might dissipate, but its net effect has been to make the atmosphere more moist, which reduces the amount of evaporation the next cloud experiences from entrainment. In summer, several such cumulus clouds might form throughout the afternoon, each one getting larger than its predecessor as the ground warms (Figure 11.2(a)).

At some point, the cumuliform cloud might become tall enough that collision and coalescence of cloud droplets within the cloud start producing precipitation, often made of big raindrops (Figure 11.2(b)). The whole process is usually relatively fast, which produces convective precipitation in the form of concentrated showers, sometimes heavy (Figure 11.3). If the cloud reaches altitudes where temperatures are below about $-10\,°C$, ice crystals form, which will be critical

for the physical processes that lead to the formation of lightning (see Section 11.3). When the cumulus cloud reaches the stable stratosphere, or a stable layer of the troposphere, it flattens out against that stable layer. (Recall that the stratosphere is very stable and that the temperature inversion prevents further upward motion.) The cloud has turned into a **cumulonimbus**, and its flattened top gives it the familiar appearance of an anvil (Figure 11.4). Strong updrafts might overshoot the level of neutral buoyancy but are quickly brought back down by the stability of the layer they overshoot into.

As precipitation falls in subsaturated air below the cloud base, it evaporates. We recall that evaporation requires latent heat that, being taken from the environment, has a cooling effect. Cold air being more dense, it tends to sink back down toward the surface, underneath the cloud. This creates cold downdrafts which compete with the convective updrafts. Friction between falling raindrops and the surrounding air also contributes to dragging the air downward and enhances the downdrafts.

When it reaches the surface, the sinking cold air forms a **cold pool** and tends to spread outward, creating a gust of cold wind that you might have experienced if you have been in the vicinity of a thunderstorm in the summer. Since full-blown thunderstorms often develop at the end of a warm afternoon, the cold temperatures accompanying the wind gust are often in sharp contrast with the summer temperatures.

When the cold pool of air becomes so extensive that it cuts off the supply of warm, humid air to the storm, the thunderstorm decays and dissipates (see Figure 11.2(c)). The whole formation and dissipation cycle takes on the order of an hour.

Meanwhile, the sinking and spreading cold air advances against warmer air, much in the same way as a cold air mass creates a cold front when it moves into a warm air mass. The boundary between the rushing cold air and the warmer surrounding air is called a **gust front**. Similarly to a cold front, it wedges underneath the warm air and forces upward motion, which can cause the formation of small cumulus clouds. In some cases, these cumulus clouds themselves grow and develop into a new thunderstorm.

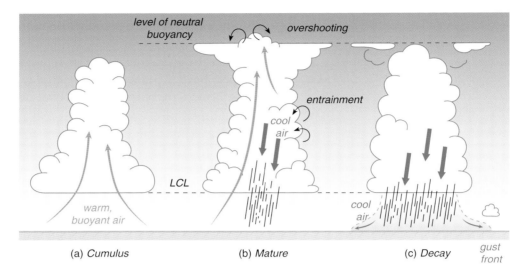

Figure 11.2. Developmental stages of an ordinary thunderstorm.

Figure 11.3. Concentrated rain shower under a growing thunderstorm.

Figure 11.4. The anvil of a cumulonimbus cloud.

It is not uncommon to observe thunderstorms developing in series by triggering each other through cold pool expansion.

11.2 Severe Thunderstorm

An ordinary thunderstorm is eventually defeated by its own downdrafts, which kill convection and remove the initial source of energy. When the cumulonimbus cloud is sheared, however, downdrafts are separated from the updrafts and can coexist symbiotically for extended periods of time, enhancing each other's intensity and the likelihood of severe weather. This can happen when the wind speed increases with height and the upper-level winds displace the top of the cloud sideways (Figure 11.5). The thunderstorm can then turn into a much more long-lived and severe weather system.

Where the anvil extends out, small precipitation particles can be blown outward with the updraft and slowly settle out of the cloud, occasionally producing round cloud pockets hanging down from the anvil. These protruding cloud elements are called **mammatus clouds**; they form as the dense air (cooled by evaporation), entrains cloud particles from the anvil downward (Figure 11.6). This is a rare example of

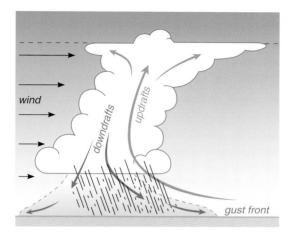

Figure 11.5. In a severe thunderstorm, the cumulonimbus is sheared horizontally and the downdrafts become separated from the updrafts.

Figure 11.6. Mammatus clouds forming on the underside of a thunderstorm anvil.

the case where subsiding air can remain saturated, because precipitation particles are available to maintain saturation through evaporation.

Cool air returns to the surface in the form of downdrafts further downstream from the updrafts, and convection can continue unhindered. Thus, severe thunderstorms can last more than 3 hours and produce extremely intense precipitation, including large raindrops, hail, as well as lightning and thunder. The wind gusts can reach 50 knots or more, and isolated, short-lived, but very intense downdrafts called **microbursts** can reach up to 100 knots. These downdrafts pose a severe hazard for aircrafts.

Since the whole system is in motion (from left to right in Figure 11.5), someone on the ground will first observe a sudden drop in temperature corresponding to the gust front. The passage of the gust front is usually also accompanied by an increase in wind speed, and possibly also a change in wind direction. After the gust front has passed, rain and possibly hail come, and possibly lightning and thunder (although these can also be evident long before the gust front, especially at night).

In summary, the conditions for the growth of a thunderstorm are:

▶ a conditionally unstable atmosphere (e.g., cold air aloft, over a warm and moist lower troposphere)

▶ a lifting mechanism (e.g., convergence along a front, or free convection forced by hot ground)

with the addition of:

▶ strong vertical wind shear

for a severe thunderstorm.

Sometimes, thunderstorms organize into a mesoscale convective system (MCS), i.e., a self-organized system made of multiple thunderstorms covering several hundred kilometers. MCSs can last several hours and produce large amounts of precipitation. New cumulonimbus clouds develop along the leading edge of the cold pool created by the downdrafts and the sinking cold air, sustaining the MCS. As the cold pool expands and the cold air displaces warmer air aloft, a **shelf cloud** can form over the gust front (Figure 11.7).

When convection is triggered by an advancing cold front, as often happens in the southern and eastern United States, thunderstorms may become organized along the front and form a **frontal squall line**, provided that wind shear is again available to separate the updrafts and downdrafts. They are easily identified in infrared satellite imagery as a thin line of tall (white) clouds, typically along the tail of a cold front associated with a midlatitude cyclone, and in radar imagery as a line of heavy precipitation (see Figure 2.12).

Figure 11.7. Shelf cloud forming over an advancing gust front.

11.3 Lightning and Thunder

Cumulonimbus are often associated with lightning and thunder, when convection is violent and precipitation intense. Lightning forms when an electric field builds up an electric *differential* that exceeds the capacity of the air, i.e., an electric charge difference between two regions of the atmosphere that grows so large that a violent discharge occurs in the form of a spark (lightning), which reduces the electrical charge difference. Lightning can occur between two clouds, between parts of a cloud (in which case the cloud seems to light up from inside), or between a cloud and the ground, which is the type of lightning stroke we usually observe, because it is clearly visible and it can impact us directly – even though cloud-to-ground strokes account for less than 20% of all lightning strokes.

Charge separation occurs when larger ice particles, such as graupel and hailstones, fall through the cloud and collide with smaller ice particles ascending in the updrafts. The larger falling particles become negatively charged while the smaller lofted particles become positively charged. This tends to concentrate negative charges in the lower part of the cloud and positive charges in the upper part of the cloud (Figure 11.8(a)). Comparatively, the ground is charged positively. And when the charge difference between the cloud and the ground becomes too great

for the air to sustain, negative charges start moving downward toward the ground.

They do not do so all at once, however. They penetrate through the air in successive pokes, advancing ever further down along a path, extending the path by a discrete step at each new poke. This advancing discharge is called a **stepped leader** and it moves extremely rapidly – on the order of microseconds – choosing the path of least resistance (Figure 11.8(b)). As it approaches the ground, it induces a concentration of positive charges in the "closest" elements of the landscape, i.e., elements that stand out and reach up in the atmosphere, in particular pointy objects such as trees, electric poles, and metallic objects such as found at the top of church steeples (Figure 11.8(c)). That is why it is recommended to stay away from trees during a thunderstorm.

The positive charges eventually move up from these protruding elements to meet the stepped leader, connecting and opening the path for further electrons to flow down: this subsequent flow of electrons is the real lightning stroke, a much larger discharge of electricity, called the **return stroke**, that heats up the air to an incredible 30 000 K and shows as a bright white path of light (Figure 11.8(d)). [Recall the relationship between temperature and the emission of radiation – here, the emission of visibe light by heated air molecules.]

The rapid increase in temperature in the narrow channel of air is accompanied by a sudden increase in pressure. In other circumstances, the air would adjust by expanding as the pressure increases, as air parcels do when they rise or subside in the atmosphere. The discharge of electricity is so violent, however, and the build-up of pressure so rapid, that the channel expansion creates a shock wave. The resulting sound wave travels outward from the channel, and we eventually hear it as **thunder**.

Because the flash of light created by the return stroke travels at the speed of light (i.e., 300 000 000 m/s), we see it almost instantaneously. The sound wave, however, travels at the speed of sound, which is closer to 340 m/s, or about one kilometer in three seconds (i.e., about a mile in five seconds). Therefore, we usually hear thunder some time after we see the return stroke, and, by counting how many seconds have elapsed

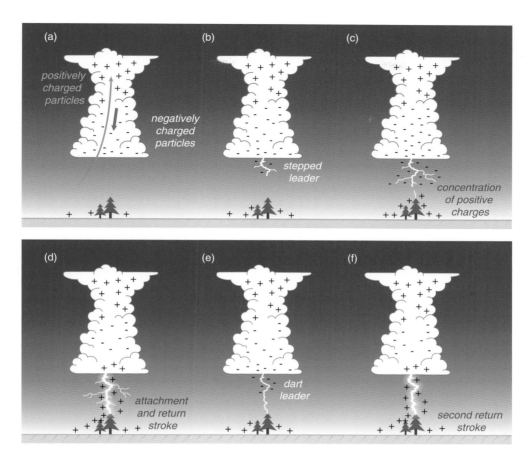

Figure 11.8. A sequence of steps leading to cloud-to-ground lightning.

between the two of them, we can estimate the distance between us and the cumulonimbus. For example, if 6 s have elapsed, we know the thunderstorm is 2 km away. If, some time later, a new stroke produces thunder only 5 s later, we know that the thunderstorm, or at least the lightning-producing part of the cloud system, is in motion, and it is approaching.

Our lightning story does not end with the return stroke. Once the channel is established, more electricity can flow through it, if the cloud still contains pockets of negative charges (Figure 11.8(e)). The new flow of electrons traveling the established channel is called the **dart leader** and causes more return strokes, all happening in less than a second. That is why lightning often appears to flash very rapidly and repeatedly when it occurs. These new return strokes all borrow the main channel, the one path that was selected when it attached to a positively charged element at

the surface (Figure 11.8(f)). That is at the expense of other paths that did not connect to the ground – failed attempts, as it were, as the stepped leader was probing and penetrating down through the air. These branches can be seen in Figure 11.9, surrounding the much brighter channel traveled by the return stroke.

11.4 Supercells

In certain circumstances, intense **supercell** storms can develop. Supercells are long-lived thunderstorms with rotating updrafts, sometimes embedded in a larger weather system, such as a midlatitude cyclone. Supercells might be triggered and sustained by an advancing cold front, for example. The rotating dynamics induce a lower pressure core within the updraft called a *mesocyclone*.

Figure 11.9. Bright white return strokes from three different areas of lightning discharge.

Supercells acquire their rotation from the vertical wind shear associated with the increase in wind speed, and/or change in wind direction, with height. You can convince yourself that wind shear creates rotation by imagining a spinning wheel embedded in the wind shear, depicted schematically in Figure 11.10. Since the wind is blowing faster at the top than at the bottom of the wheel, the wheel will start spinning clockwise (in our example). By a slightly bigger stretch of imagination, we can imagine air parcels spinning clockwise, and in fact entire horizontal "rolls" or tubes of air (going into the page) spinning clockwise, as if rolling a carpet (Figure 11.11(a)). So far, these rolls are parallel to the ground, but, in the presence of a strong convective updraft, parts of these tubes are pushed upward, distorting the roll as shown in Figure 11.11(b). Two vertical segments appear on each side of the convective center, spinning in opposite directions.

The fate of these two counter-rotating vortices is better captured in a bird's eye view of the evolving convective cell (Figure 11.12). The cell splits into two individual rotating systems. The anticyclonic vortex, moving to the left, commonly weakens and decays, while the cyclonic vortex moves to the right and intensifies, due to an enhanced buoyancy associated with the development of low pressure in the right-moving

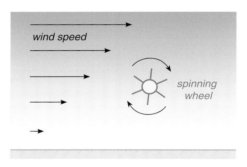

Figure 11.10. Vertical wind shear creates horizontal rotation.

part of the storm. In fact, this part of the storm turns into a mesocyclone characterized by a well-defined low pressure center.

As this new system intensifies, updrafts and downdrafts become organized (Figure 11.13). As warm, moist air swirls into the convective cell to form the rotating updraft that, counter to intuition, is so strong that it is relatively free of precipitation, downdrafts bring cool, dense air back down to the surface, along with precipitation, at the front and rear of the cell. The cool air spreads out and forms two gust fronts, on the rear and forward flanks of the convective cell. Rotation can develop very near to the ground as a result of the outflow spreading

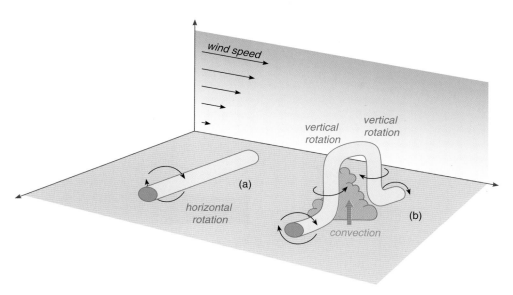

Figure 11.11. Transformation of horizontal rotation (a) into vertical rotation (b) in the form of two counter-rotating vortices, as convection distorts the rotating tube.

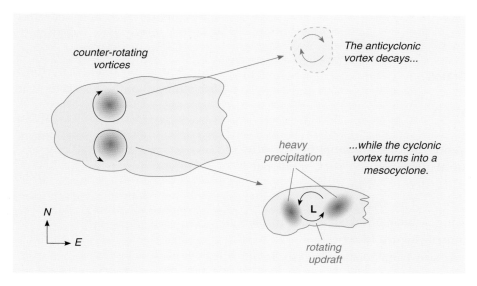

Figure 11.12. Convective cell splitting into a decaying anticyclonic vortex and a growing mesocyclone.

beneath the updraft, and occasionally the rotation can be intensified to form a **tornado**, a fast-spinning vortex of much smaller scale than the thunderstorm itself. This is most likely to happen when the outflow is not excessively cold, and when the overlying thunderstorm updraft is particularly intense within about 1 km of the surface.

Within the rear-flank downdraft there often appears an area of precipitation that wraps around the incoming warm air and takes the form of a *hook* in radar images. Meteorologists use the formation of a hook in radar imagery as an indicator of the location where a tornado might form. While hook echoes are a good indicator of rotation associated with a

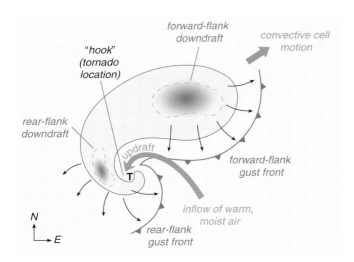

Figure 11.13. Structure of a tornadic supercell, with preferred location (the "hook") for the formation of a tornado (**T**).

Figure 11.14. Dirt and debris stirred by an F3 tornado.

supercell, only a subset of supercells produce tornadoes (roughly 20%).

11.5 Tornadoes

11.5.1 Description

Tornadoes are rapidly rotating columns of air with extremely low central pressure that suck small particles, dust, sand, dirt, debris, and sometimes larger objects into their core. The debris rises and spins around with the wind, which often gives the tornadoes a dark appearance (Figure 11.14).

Tornadoes vary in size from tens to hundreds of meters, but are always much smaller than a thunderstorm. They are primarily spawned by supercells. The intensity of a tornado is determined by the strength of its winds using a scale designed by Tetsuya Fujita in the 1970s (see Box 11.1). Tornadic winds can reach more than 250 knots. The tornado is normally a single vortex, but may also spawn suction vortices, which are smaller and rotate around the main vortex.

In the USA, where 75% of the world's tornadoes occur, the strongest tornadoes tend to occur from April to June, and in so-called *Tornado Alley* (see Section 11.5.3).

Box 11.1.   The Fujita scale

Tetsuya Fujita (1920–1998) designed a tornado intensity scale based on damage surveys in the 1970s that was soon adopted by the entire atmospheric science community (and was recently replaced by the Enhanced Fujita Scale). He ranked the intensity of tornadoes based on the destruction caused by the winds and using the strength of the winds as follows:

F0	35–62 knots
F1	63–97 knots
F2	98–136 knots
F3	137–179 knots
F4	180–226 knots
F5	227–276 + knots

11.5.2 Tornado Development

A precursor to the tornado is usually the formation of a **wall cloud** hanging from the cloud base and slowly rotating around the mesocyclone (Figure 11.15). The wall cloud is lower than the cloud base because the air entering the wall cloud is cool and humid. A **funnel cloud** then emerges from the wall cloud or the base of the thunderstorm (Figure 11.16). As the inflow of air accelerates by conservation of angular momentum and the vortex spins up – recall the ice skater spinning faster as she pulls her arms and legs inward – the

Figure 11.15. Wall cloud hanging below the cloud base of a thunderstorm.

Figure 11.17. The vortex spins up and turns into a full-blown tornado.

Figure 11.16. Funnel cloud nearly reaching the ground beneath a supercell thunderstorm.

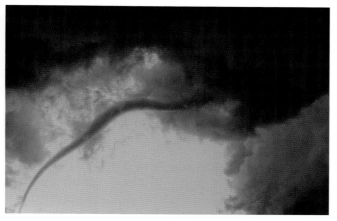

Figure 11.18. Stretching and distorsion of a tornado.

pressure decreases inside the vortex and the funnel cloud extends downward, sometimes down to the surface (Figure 11.17). Meanwhile, the air is rising rapidly in the vortex, due to intense "suction" from above, near the cloud base. The vortex is typically elongated between the base of the cloud and the ground. This stretching concentrates the rotation (technically, angular momentum), much like water approaching a bathtub drain can spin up a stronger vortex. Thus, the tornado can continue to intensify for minutes, and sometimes tens of minutes, all the while moving with the convective cloud, leaving a swath of destruction in its wake.

Eventually, the overlying storm updraft weakens, or the tornado is displaced laterally from the overlying updraft, which reduces the "suction" from above, and therefore reduces upward motion and the convergence of angular momentum. Alternatively, the air feeding the tornado can become too cold, which reduces buoyancy and upward motion. During the dissipation phase, tornadoes are sometimes stretched into a long, thin cloud "rope" (Figure 11.18). For lack of a rotating updraft and warm, moist air inflow, the vortex eventually dissipates, but the supercell can spawn more tornadoes further down its path, giving birth to a *tornado family*. Some tornadoes can grow in radius and then give birth to suction vortices – smaller, but incredibly fast-spinning, vortices with winds up to 250 knots – that form on the outskirts of the main tornado.

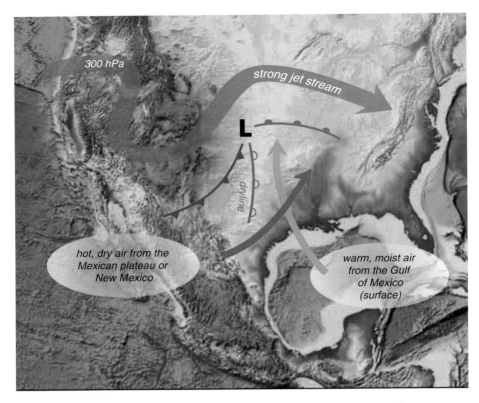

Figure 11.19. Synoptic situation conducive to the formation of supercells and tornadoes.

Scientists have a pretty good understanding of the fact that the convergence of angular momentum is needed to make a tornado. This, generally, is most likely to happen when the outflow is not too cold and the wind shear in the environment is very strong, because this shear promotes the "suction" effect that is key to pulling cool air upward and angular momentum inward near the surface. Scientists, however, are not yet able to predict if/when these processes will suddenly occur within a supercell storm, or what other processes might accelerate tornado formation.

11.5.3 Tornado Alley

Why do tornadoes occur in the US Midwest more than anywhere else in the world? The Great Plains are a wide open corridor for air masses to circulate and clash. Cold air masses from Canada meet warm, moist air masses from the Gulf of Mexico and form mid-latitude cyclones. These cyclones are associated with upper-level troughs in the jet stream, which provide a lifting mechanism to bring air to the level of free convection (LFC; see Chapter 6). Furthermore, lifting can be enhanced along boundaries such as a cold front, which provides a lifting mechanism to trigger convection and the formation of thunderstorms in the warm, unstable air ahead (Figure 11.19). Sometimes, the lifting is provided by the boundary between continental dry air advancing against moist air from the Gulf – a boundary called a **dryline**. A strong upper-level jet provides the shear necessary to create horizontal rotation, which is turned into vertical rotation by convection, as we explained earlier (Figure 11.11). We have all the ingredients for a supercell, and eventually the possibility of tornadoes.

But the region provides yet another crucial ingredient. Hot, dry air from the Mexican plateau flows into the basin and flows over the warm, moist air from the Gulf of Mexico. This creates a temperature inversion right over the inflow of warm and moist air, at about 850 hPa. Figure 11.20(a) shows how buoyant air parcels lifted from the surface (blue arrow) cannot rise

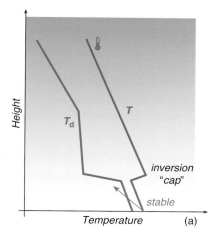
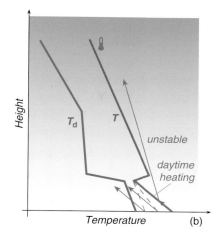

Figure 11.20. Typical temperature profiles in (a) the morning, as the capping inversion first prevents convection and (b) in the afternoon, after daytime heating allows buoyant air parcels to overcome the inversion.

higher than the inversion, at first. As the lower atmosphere warms up, however, due to daytime heating, the temperature of the buoyant parcels increases, so that at some point, they can rise up to 850 hPa and still be warmer than the maximum temperature of the inversion (Figure 11.18(b)). Being still buoyant, they can continue their ascent, and a cumulonimbus cloud can form. Most importantly, while the lower atmosphere was slowly heating during the day, it was storing a lot of potential energy, which is the main role of the capping inversion. A good analogy is a pressure cooker in which the temperature and pressure increase, but the air can neither expand nor rise. If we remove the lid, the air will rush out with tremendous speed and energy. Another analogy could be a car driver pressing on the accelerator of the car while holding the brake pedal down at the same time. As soon as the brake pedal is released, the car experiences a huge acceleration. Similarly, from the moment the inversion cap is "punctured," air parcels accelerate rapidly due to the very large buoyancy force, and the supercell thunderstorm intensifies very rapidly.

Our understanding of tornado dynamics has greatly improved with the installation of a network of Doppler radars and the deployment of mobile Doppler radars during dedicated field campaigns. The exact location and timing of tornado formation remain difficult to forecast, but the factors that are conducive to tornado formation can be assessed with observations and numerical weather prediction models. Forecasters routinely estimate the stability of the atmosphere, the level of free convection, and the potential for violent thunderstorms to develop. They estimate the vertical wind shear in critical areas where supercells are known to form, such as along frontal boundaries. And they detect the presence of capping inversions in vertical soundings. When the conditions are ripe, forecasters concentrate on Doppler radar imagery to identify the presence of a *hook echo* and any cyclonic rotation that could be the signature of a mesocyclone. When a tornado is forming, the weather service may have a few minutes to issue a tornado warning for the local population to seek shelter.

Summary

Thunderstorms form by convection in an unstable or conditionally unstable atmosphere, with varying degrees of intensity and severity depending on the environmental conditions.

Ordinary thunderstorms grow in the absence of wind shear. Cold downdrafts created by evaporational cooling cause the formation of a **cold pool** that tends to spread outward and to create a gust of cold wind that advances against warmer air. The resulting boundary, or **gust front**, can spawn new convective clouds and new thunderstorms.

When the cold pool of air becomes so extensive that it cuts off the supply of warm, humid inflow to the storm, the thunderstorm decays and dissipates.

In the presence of vertical wind shear, updrafts and downdrafts can coexist symbiotically, and the storm can be long lasting and **severe**, with intense precipitation and strong isolated downdrafts called **microbursts**.

Long-lived thunderstorms with rotating updrafts, called **supercell thunderstorms**, can induce a lower pressure core within the updraft called a **mesocyclone**. They acquire their rotation when horizontal rotation associated with vertical wind shear is reoriented by convection. In some supercells, a much smaller region of rotation can develop very near to the ground and, occasionally, intensify to form a **tornado** – a fast-spinning vortex of much smaller scale than the thunderstorm itself.

The American Midwest is a prime location for tornado formation due to the absence of barrier for cold air masses from Canada and warm air masses from the Gulf of Mexico. The air mass encounter provides all the necessary ingredients for tornado formation:

▶ lifting by a cold front or a dryline to trigger convection and the formation of thunderstorms in the warm, unstable air;

▶ vertical wind shear provided by the upper-level jet, to create horizontal rotation, later turned into vertical rotation by convection; and

▶ inflow of hot, dry air from the Mexican plateau, which creates a temperature inversion above the inflow of warm and moist air from the Gulf of Mexico. This allows the surface air to store convective potential energy before the inversion is punctured, and the buoyant air swiftly rises to form a cumulonimbus.

This region of the USA, referred to as "tornado alley," is carefully monitored by weather forecasters, who use observations, Doppler radars, and numerical weather prediction models to detect the factors and conditions leading to the formation of tornadoes, and warn local populations in time.

CHAPTER 12

Tropical Cyclones

Formidable heat engines of the tropics, tropical cyclones garner enormous quantities of energy from the ocean and unleash some of the most destructive winds on Earth. They confront us with some of the most compelling dynamics of the atmosphere and still challenge our weather prediction systems. In this chapter, we describe the structure and development of tropical cyclones, and explain the mechanisms and factors that contribute to their intensification, or demise.

CONTENTS

Tropical cyclones are the most powerful and the most destructive weather systems produced by the atmosphere. Years later, local populations still remember and talk about Hurricane Andrew in Florida in 1992 (thousands of destroyed homes, 65 dead, and billions of dollars of damage) and Hurricane Mitch in Central America in 1998 (record-breaking rainfall, catastrophic flooding, 11 000 killed). Typhoons and tropical cyclones cause even more damage in Asia. We often cite the historical tropical cyclone that killed 500 000 in Bangladesh in 1970, but more recently, and in spite of all the technological advances in forecasting typhoons and warning local populations, Super Typhoon Haiyan killed 6300 in the Philippines alone in 2013.

Before we try to understand how tropical cyclones come about and develop, let us gather some facts and figures about them from the different observation systems available to us.

12.1 Facts and Figures

Tropical cyclones are tropical storms with sustained winds greater than 64 knots. [Recall that, if the wind is 64 knots when averaged over one minute, wind

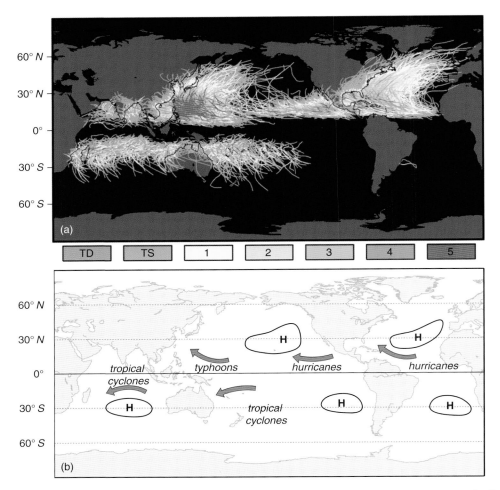

Figure 12.1. (a) About 150 years of hurricane tracks, color coded by intensity according to the Saffir–Simpson hurricane intensity scale (tropical depression (TD), tropical storm (TS), and category 1–5). (b) Overall direction of motion of the different tropical cyclones, along with the names used in each ocean basin.

gusts can be much higher.] We tend to use the term **hurricane** when they form over the Atlantic Ocean or Eastern Pacific Ocean, **typhoon** when they form over the Western Pacific Ocean, and **tropical cyclone** when they form over the Indian Ocean, although they all share the same dynamics (Figure 12.1). They are given names to facilitate discussion (e.g., among weather forecasters) as well as research, because atmospheric scientists often refer to old cases to illustrate or investigate specific aspects of tropical cyclone development. A list of names is created for each ocean basin, starting with the letter A and alternating male and female names in alphabetical order

(e.g., Arlene, Bret, Cindy, etc. for 2017 Atlantic hurricanes). When a tropical cyclone is particularly devastating and becomes a reference case in the scientific literature, its name is "retired" from future lists, so that it becomes uniquely associated with that historical event. Such is the case of Hurricanes Katrina and Rita, for example, during the record-breaking 2005 Atlantic hurricane season.

Tropical cyclones are ranked based on maximum sustained wind speed, first as tropical disturbance, tropical depression, and tropical storm up to 64 knots, and then by category 1 to 5 according to the Saffir–Simpson scale for hurricane intensity (see Table 12.1, where

Table 12.1. *The Saffir–Simpson hurricane intensity scale*

	Central pressure (hPa)	Winds (knots)
Tropical disturbance		<20
Tropical depression		20–34
Tropical storm		34–64
Cat. 1	>980	64–82
Cat. 2	965–979	83–95
Cat. 3	945–964	96–113
Cat. 4	920–944	114–135
Cat. 5	<920	>135

central pressures are indicated for reference, based on an algorithm). Category 5 tropical cyclones are the most intense, but are relatively rare – in fact, before Hurricane Irma in 2017, the last category 5 hurricane in the Atlantic basin was Hurricane Felix in 2007.

Tropical cyclones have a low pressure center, much in the same way as midlatitude cyclones, although their dynamics are very different. The lowest central pressure ever recorded is 870 hPa in Typhoon Tip in 1979. Cyclonic winds blow around the low pressure center (clockwise in the southern hemisphere and counterclockwise in the northern hemisphere), which suggests that the Coriolis force plays a role in explaining the formation and development of tropical cyclones – a point to remember.

Although tropical cyclones share some similarities with midlatitude cyclones, they differ in many important ways. Tropical cyclones are about 300 to 700 km in diameter, and therefore smaller than midlatitude cyclones. They have no fronts, and are more symmetric than midlatitude cyclones. They tend to last longer, to have stronger maximum winds, and to cause more damage. Their central core is warmer than their surroundings, i.e., the opposite of midlatitude cyclones, which usually have a cold core. Indeed, tropical cyclones and midlatitude cyclones have very different dynamics and, in particular, draw their energy from

different sources. Midlatitude cyclones draw their energy from the pole-to-tropics temperature gradient, whereas tropical cyclones draw their energy from the underlying ocean, as we shall see shortly.

A map summarizing the geographic distribution of tropical cyclones (Figure 12.1) tells us that they occur mostly between about 5° and 25° of latitude. Figure 12.1 also shows us a number of interesting features:

▶ Tropical cyclones form over the ocean, and not over land. In fact, they either die or turn into a midlatitude cyclone after making landfall.

▶ Tropical cyclones clearly do not form at the equator, nor do they form in the midlatitudes.

▶ Tropical cyclones rarely form in the South Atlantic and do not form in the Eastern South Pacific.

▶ Although their individual tracks can be somewhat erratic (see Figure 12.2, for example), tropical cyclones follow a mostly westward trajectory, and sometimes curve poleward before dying or turning into a midlatitude cyclone. They seem to follow the flanks and curve around the subtropical highs.

▶ The Western Pacific Ocean experiences the most intense tropical cyclones, and in particular many category 5 storms.

A comparison of the 2005 Atlantic hurricane tracks with ocean temperature at the beginning of the season (officially, June 1) in Figure 12.2 suggests that tropical cyclones tend to form and strengthen where the ocean is warm, a good hint that tropical cyclones are indeed using energy drawn from the ocean to develop and grow. Note how the four tropical cyclones that reached category 5 that year (a record-breaking hurricane season by all measures), Emily, Katrina, Rita, and Wilma, all reached their highest intensity over the Caribbean and the Gulf of Mexico, where ocean waters are the warmest.

Before we make sense of all these observations, let us observe some tropical cyclones from space in more detail.

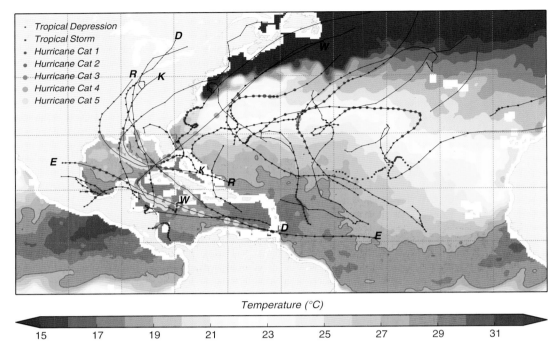

Temperature (°C)

15   17   19   21   23   25   27   29   31

Figure 12.2. Hurricane tracks of the record-breaking 2005 Atlantic hurricane season. The size and color of the dots indicate hurricane intensity. The start and end points of Hurricanes Dennis (*D*), Emily (*E*), Katrina (*K*), Rita (*R*), and Wilma (*W*) are indicated. The background temperature field is constructed from AMSR-E sea-surface temperature measurements on June 1, 2005. The red contour corresponds to 28 °C.

12.2 Tropical Cyclone Structure

What can we learn from satellite imagery? In the example shown in Figure 12.3, we can see that the top of a tropical cyclone is made of cirrus clouds punctured by tall convective clouds: cumulonimbus clouds corresponding to thunderstorms. These thunderstorms are organized in spiraling bands that circle around the **eye**, a tight area at the center of the storm, 40 to 60 km in diameter, that is often free of high clouds. The eye is surrounded by a wall of intense convection called the **eyewall**, where the strongest winds and heaviest rain are found.

These observations hint at how tropical cyclones come about and grow. They are essentially made of spiraling convective clouds drawing their energy from the underlying warm ocean. As the converging surface winds evaporate water from the ocean, latent heat is taken from the ocean, focused into the core of the tropical cyclone, and carried upward by convection (Figure 12.4). As the water vapor condenses higher up to form cloud droplets and raindrops, the latent heat is released and warms the core of the tropical cyclone, which lowers the surface pressure. In turn, the lower pressure pulls even more surface air into the core of the tropical cyclone, accelerating the surface winds and the evaporation of water. We are back to step one and have a positive feedback loop that continues as long as the tropical cyclone is in a favorable environment over warm ocean water and until it reaches its maximum intensity. Therefore, a tropical cyclone over water is a self-sustained heat engine that keeps intensifying as long as the environmental conditions are favorable.

Aloft, as the air reaches the tropopause, it spreads outward in the form of extensive cirrus clouds, because it cannot penetrate into the very stable stratosphere due to the temperature inversion (similar to the anvil top of a thunderstorm). It then subsides further out, around the tropical cyclone, which explains why the surroundings of a tropical cyclone are usually free of low clouds. (Recall that convection causes

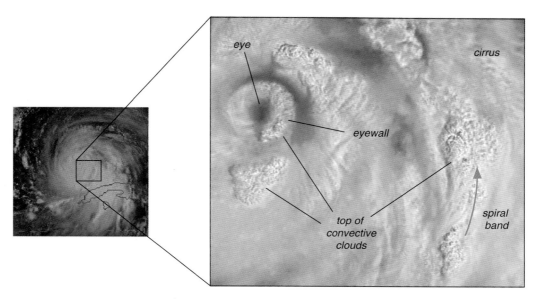

Figure 12.3. Satellite view of Hurricane Katrina on August 27, 2005. A close-up of the eye reveals spiraling bands of cumulonimbus clouds.

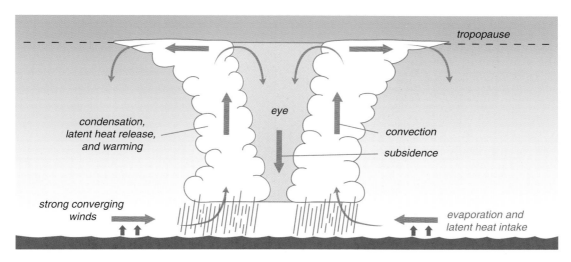

Figure 12.4. Structure and transversal wind circulation in a tropical cyclone.

cloud formation, while subsidence *inhibits* cloud formation).

While most of the upper air flow is directed outward, some air returns down into the eye, where it subsides, which explains why the eye is mostly free of mid- and high-level clouds. Moreover, subsiding air warms due to adiabatic compression, which explains why the core of a tropical cyclone is exceptionally warm and comparatively dry (i.e., low relative humidity).

This basic mechanism provides a good indication as to why tropical cyclones can form only over the ocean,

and why they tend to form over *warm* oceans: they are essentially fueled by latent heat release in the convective towers. The latent heat originates from evaporation of warm ocean water by the strong surface winds. Evaporation of warm water is crucial, and the tropical cyclone quickly dies as soon as it finds itself over land, or over cooler water. This explains why tropical cyclones do not form over the eastern South Pacific ocean: the water is too cold.

If we apply the same reasoning as we did for the sea breeze circulation, we find that warming from

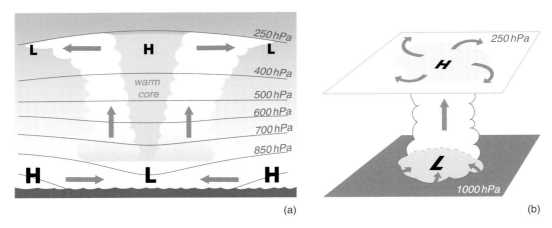

Figure 12.5. (a) Pressure distribution and (b) corresponding surface and upper-level circulation in a tropical cyclone.

latent heat release in the core of the tropical cyclone also expands the central air column and lowers the surface central pressure (see Chapter 8). This gives a pressure distribution such as shown in Figure 12.5. Surface pressure is lowest in the eye of the tropical cyclone, and surface winds blow cyclonically around the low while converging into the core of the tropical cyclone, due to surface friction (Figure 12.6). At the tropopause, however, pressure is highest in the center and drives upper-level winds outward. As they diverge away, the upper-level winds create a spiraling cirrus top easily identifiable on satellite imagery. The two circulations are well captured in Figure 12.7, where we can contrast the converging streamlines at the surface (Figure 12.7(a)) with the diverging streamlines at 250 hPa, i.e., at the tropopause (Figure 12.7(b)).

Surface winds tend to accelerate as they converge inward, by partially conserving angular momentum. Recall the ice skater spinning faster as she pulls her arms in, and the surface winds accelerating in a mid-latitude cyclone as they converge into the low pressure center. Similarly, in a tropical cyclone, the strongest winds are found close to the eye (see the blue ring around the eye in Figures 12.6 and 12.7(a), which is the location where inward spiraling air stops moving toward the eye and abruptly rises into the eyewall).

Thus, if we were to experience the passage of a tropical cyclone, we would observe pressure and wind speed changes such as shown in Figure 12.8

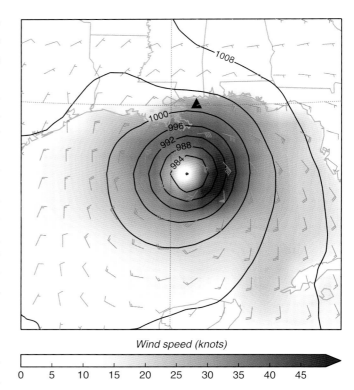

Wind speed (knots)

Figure 12.6. Surface pressure and winds on August 29, 2005, as Hurricane Katrina approaches New Orleans, Louisiana. The black triangle shows the location of NDBC buoy 42007. [The pressure gradients and winds are an output from a numerical weather prediction model at intermediate resolution and are greatly underestimated.]

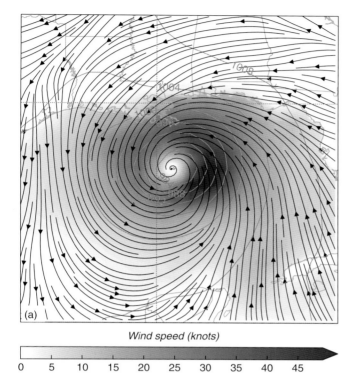

Wind speed (knots)

Figure 12.7. (a) Surface streamlines overlaid on surface pressure and wind speed on August 29, 2005. (b) Corresponding 250 hPa streamlines overlaid on GOES-12 infrared satellite imagery.

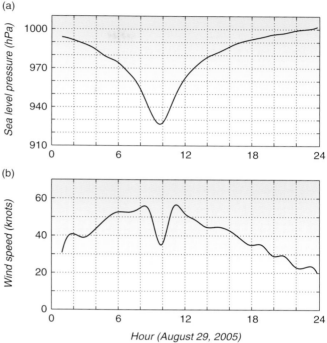

Hour (August 29, 2005)

Figure 12.8. (a) Pressure and (b) wind speed observations as Hurricane Katrina passes over NDBC buoy 42007 on August 29, 2005.

for buoy 42007 from the National Data Buoy Center (NDBC) at the National Oceanic and Atmospheric Administration (NOAA). The location of buoy 42007 is indicated with a black triangle in Figure 12.6 for reference. It is anchored off the coast of Mississippi, east of New Orleans, along Hurricane Katrina's path before making landfall. To understand Figure 12.8, imagine that the pressure and wind fields shown in Figure 12.6 are shifting northward over the black triangle. As the hurricane approaches, the buoy first recorded a decrease in pressure down to a minimum of 927 hPa, followed by an increase in pressure as the storm continued northward. Consistent with these observed changes in pressure, the wind speed first increased to a maximum of 54 knots as the ring of strongest winds around the eye approached, and then decreased as the storm progressed northward. Note, however, the "lull" around 11:00 a.m. when the buoy was close to the eye of Katrina and was experiencing relatively weaker winds for an hour or so.

The vertical structure of the tropical cyclone shown in Figure 12.5 suggests that the vertical "stacking" of surface low pressure and upper-level high pressure in the warm core is necessary for the eyewall to exist and the overall circulation to continue. Contrary to thunderstorms, which become more severe weather systems when they are sheared and tilted, allowing downdrafts to become separated from updrafts, tropical cyclones weaken when there is a strong wind shear with height. That is why tropical cyclones only form and grow in the tropics, where vertical wind shear is weak, and not in the midlatitudes, where the presence of the upper-level jet stream implies a strong vertical wind shear. In fact, when tropical cyclones venture into the midlatitudes, their structure changes and they either dissipate or turn into a midlatitude cyclone.

12.3 Tropical Cyclone Development

Now that we have a good picture of tropical cyclone structure at the mature stage, let us explore the mechanism that causes them to form in the first place. A natural starting point involves retracing the steps of a hurricane back to the eastern side of the Atlantic Ocean basin.

12.3.1 Tropical Easterly Wave

We recall that the Intertropical Convergence Zone (ITCZ) is a region of pronounced convection, where the converging trade winds force upward motion and the formation of tall cumulonimbus clouds. Most of the time, these thunderstorms are disorganized. They grow and dissipate, following the easterly flow for a few hours or days. However, the subtropical surface easterly flow is wavy and contains troughs and ridges, much in the same way as the midlatitude jet stream. In the Atlantic Ocean, for example, the undulations of the easterly flow form over the Ethiopian Highlands of East Africa, before moving westward across North Africa.

We also recall that the wind tends to be subgeostrophic around troughs and supergeostrophic around

ridges (see Box 8.4 and the blue arrows in Figure 12.9). As a result, the easterly flow contains regions of convergence and regions of divergence, on either side of the trough axis. Compared to troughs in midlatitudes, notice that the pressure gradient is reversed: whereas pressure overall decreases northward *at upper level* in the northern midlatitudes, from subtropical high pressures toward the polar trough (see Figure 9.1), pressure overall *increases* from the equator to the subtropical highs *at the surface* (see Figures 9.8 and 9.11). Therefore, the trough pattern is rotated by 180°.

Thunderstorms tend to congregate east of a trough axis, where convergence enhances convection. These convective patches can be identified on satellite imagery and are usually followed closely by weather forecasters, because they can later develop into a stronger system, and, for a small subset, into a hurricane. At this stage, we refer to this early precursor as a **tropical easterly wave**. Such a wave appears in Figure 12.10 off the coast of Africa on September 1, 2009. The intense wave and cloud activity further upstream over Africa will later evolve into another easterly wave and Hurricane Fred over the eastern Atlantic.

12.3.2 Tropical Depression

We recall that, in a thunderstorm (i.e., in a cumulonimbus), the updrafts compete with downdrafts bringing cold air down to the surface. These downdrafts tend to stabilize the lower atmosphere and to suppress further development. As the cold and dry air comes in contact with the ocean surface, it removes heat the lower atmosphere had gained from the upper ocean, and the whole convective process has to start again with warming and moistening the air near the surface. For a cluster of cumulonimbus to evolve into a tropical depression, convection needs to persist so as to keep warming and moistening the lower and middle troposphere. As more and more heating occurs, the updrafts start winning over the cold downdrafts, and the net heating becomes overall positive and penetrates deeper into the troposphere. Each new convective cell transports more moisture aloft, and the heating accelerates. This process is enhanced

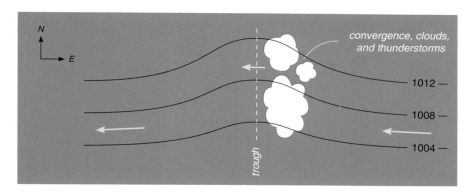

Figure 12.9. Tropical easterly wave embedded in the surface trade winds.

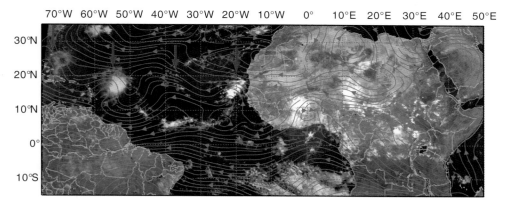

Figure 12.10. Visible satellite imagery and 700 hPa streamlines on September 1, 2009, at 12:00 UTC, with three easterly waves indicated by red arrows.

when the air in the middle troposphere moves with the region of convection; that is, when the moistening middle tropospheric air stays with the region of convection. This tends to happen when vertical wind shear is weak; strong shear tends to introduce dry air continually into the middle troposphere, which enhances the formation of downdrafts.

As shown in Figure 12.5, net heating of the air column lowers the surface pressure. Surface winds start to converge and to accelerate by partial conservation of angular momentum, which enhances evaporation, the transport of moisture into the forming warm core, convection, etc.: our positive feedback loop is initiated. The warming also raises the pressure surfaces and creates an upper-level high (see Figure 12.5 again), which forces air to flow out of the area, consistent with lowering the surface pressure. With lower pressure near the surface, rotation

increases under the influence of the Coriolis force. At this point, the disorganized convection has turned into an organized **tropical depression** (TD), with sustained winds of 20 to 34 knots.

An example of a tropical depression can be seen in Figure 12.10 east of the Caribbean Islands, along with the corresponding easterly wave. It evolved into Tropical Storm Erika over the following days.

12.3.3 Tropical Storm

With a continuation of the processes described in Section 12.3.2, the system intensifies and turns into a **tropical storm** (TS) with sustained winds of 34 to 64 knots. It is steered westward by the overall easterly flow, i.e., the southern flank of the subtropical high (see Figure 9.13).

12.3.4 Tropical Cyclone (Hurricane)

As the storm intensifies, an eye forms and becomes well defined, with an eyewall that concentrates the strongest convection and rainfall. The storm technically becomes a **tropical cyclone** when sustained winds reach 64 knots. It is then ranked by category from 1 to 5 according to the Saffir–Simpson scale for hurricane intensity (Table 12.1). Tropical cyclones typically grow in intensity as they move westward and cross the ocean basin, but they can also decrease in intensity if they encounter lower sea surface temperatures, for example, or different environmental conditions, as might be the case in the Caribbean or the Gulf of Mexico.

12.3.5 Tropical Cyclone Decay

Tropical cyclones typically weaken when they move over cooler water (lack of heat), over land (lack of moisture), or into a region of strong vertical wind shear, such as the midlatitudes. When moving over land, they usually dissipate within hours after making landfall, for lack of fuel (latent heat of evaporation), and due to friction with the land surface. This is typically the case of hurricanes encountering Central America or the southeastern USA, and typhoons encountering the Asian continent. However, the western flank of the subtropical highs has a southerly wind component that sometimes steers the hurricanes and typhoons northward (see Figure 9.13, for example, and compare it with Figure 12.1). Since ocean waters are usually colder at higher latitude, northward motion often announces the demise of the storm. Yet, when the tropical cyclone starts drifting poleward and its position is coincident with an approaching upper-level trough in the midlatitude jet stream, the tropical cyclone can morph into a midlatitude cyclone. It then develops a warm sector, cold and warm fronts, and its dynamics change to resemble those of midlatitude weather systems.

Tropical cyclones cause a lot of damage in different ways. Extremely high winds can destroy poorly built structures, damage the electric grid, and topple trees over a large area. Extreme rainfall often saturates the soil and causes extensive flooding. Landslide can

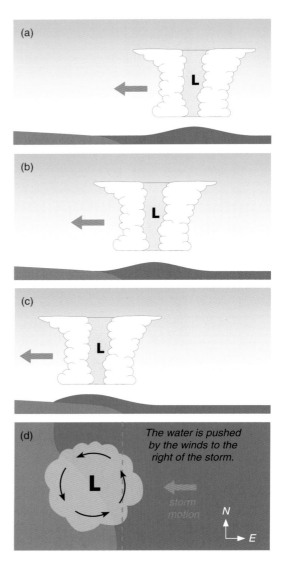

Figure 12.11. (a–c) Storm surge propagating westward under a tropical cyclone. (d) Bird's eye view showing the winds to the right of the storm pushing the surge water against the coastline in the direction of motion of the storm.

ensue, which causes more problems, including failure of the electrical grid, breaking of water pipes, roads and houses. And the destruction of infrastructure can have unexpected consequences, such as even more flooding, the absence of potable water for days or weeks, and the spread of diseases.

One of the more dangerous aspects of a tropical cyclone is the **storm surge**, which is an increase in sea level following the storm (Figure 12.11). Indeed, the low pressure at the center of the tropical cyclone allows the underlying water to *rise* above its normal level. Recall that the ocean surface is under atmospheric pressure. If you picture the atmosphere pushing more on some areas (high pressure) and less on others (low pressure), you can visualize that the ocean will be depressed under higher pressure, and bulging upward under lower pressure. Since the atmospheric pressure is extremely low at the center of a tropical cyclone, it allows the water to rise by more than one meter for larger storms. While this does not seem much in itself, it amounts to an enormous amount of water when integrated over the size of the tropical cyclone. The addition of strong winds and wave action interacting with the coastline can greatly enhance this effect to more than 8 m (26 ft), primarily to the right of the storm track where the wind is directed toward the shoreline in the northern hemisphere (Figure 12.11(d)). Strong, fast moving, and larger storms have the highest storm surge.

Storm surges often cause more damage than the wind and rainfall associated with the tropical cyclone. When they hit a flat coastal area, in particular, the rushing water can penetrate deep inland and inundate a large swath of coastal land, as was the case in Galveston Bay, Texas, in 1900 (8000 killed) and in Bangladesh in 1970 (500 000 killed). Similarly, flooding and landslides resulting from heavy tropical cyclone rainfall can kill many people, as was the case for Hurricane Mitch in 1998 (7000 killed in Honduras alone) and Super Typhoon Haiyan in 2013 (6300 killed in the Philippines alone).

12.4 Conditions for Tropical Cyclone Development

In summary, the conditions for tropical cyclone development are:

▶ A warm, conditionally unstable atmosphere. That is often true at low latitudes over a warm ocean, where there is high evaporation and a moist lower atmosphere.

▶ Warm sea surface temperatures (greater than about 27 °C).

▶ Sufficient deep ocean heat to replenish the surface as it is depleted by vertical mixing due to wind and wave action.

▶ Some rotation, and therefore some Coriolis force, i.e., tropical cyclones can only form north of about 5°N and south of about 5°S. Waves in the easterlies provide an initial source of rotation over the Atlantic Ocean, as can the monsoon trough and disturbances in the ITCZ over the Pacific Ocean.

▶ Weak vertical wind shear, i.e., tropical cyclones cannot form in the midlatitudes, where the presence of the upper-level jet stream causes strong vertical wind shear.

These conditions constrain tropical cyclones to form between about 5 and 15° latitude over the warmest sections of the tropical ocean.

The temperature of the ocean is now well monitored by satellite, and regions of high sea surface temperatures can be detected well in advance during the hurricane season. As for vertical wind shear, weather forecasters can also obtain some fairly good advance notice from the large-scale structure of the atmosphere, by observing the evolution of planetary-scale waves. Thus, even though we cannot predict the details of an evolving tropical depression, we can predict the *conditions* that are favorable to its further development into a tropical cyclone. Forecasting tropical cyclone track has improved greatly in the past 20 years. In some locations, like the North Atlantic Ocean, routine reconnaissance aircraft make detailed measurements of storm location, structure, and intensity, which complements satellite observations. These improvements in track forecasting allow the weather services to prepare local populations and to issue appropriate warnings, and allow local governments to enforce evacuation when necessary.

Summary

Tropical cyclones (hurricanes, typhoons) are tropical storms with sustained winds greater than 64 knots. Tropical systems are ranked as tropical disturbance, tropical depression, and tropical storm up to 64 knots, and then as tropical cyclone by category from 1 to 5 according to the Saffir–Simpson scale for hurricane intensity. They are about 300 to 700 km in diameter, and are composed of cumulonimbus clouds spiraling in around the central **eye**.

Tropical cyclones have a warm central core, due to latent heating associated with condensation in clouds near the eye. They draw their energy from the underlying ocean, as the converging surface winds evaporate water from the ocean (latent heat intake) and transport it into the core of the cyclone, where it condenses to form cloud droplets and raindrops (latent heat release). This warms the core of the tropical cyclone, lowers the surface pressure, and causes winds to spiral inward toward the center.

Surface winds converge cyclonically into the warm core of the tropical cyclone and accelerate until they reach the eyewall. At the tropopause, upper-level winds diverge outward, creating a spiraling cirrus top with the reverse circulation from the surface. Vertical alignment is necessary for the integrity of tropical cyclones, which explains why they cannot develop when the vertical wind shear is too strong.

In the North Atlantic, hurricanes first appear as tropical disturbances made of cumulonimbus clouds east of the axis of troughs, where convergence enhances convection – early precursors called **tropical easterly waves**. For a small fraction of easterly waves, convection is able to moisten the atmosphere sufficiently so that latent heating warms the air in the system, resulting in lower pressure. This in turn enhances convergence of air into the low pressure, which accelerates the wind, forming a **tropical depression**. This process may continue, with the storm intensifying into a **tropical storm**, and possibly even a **tropical cyclone**, when the eye often first appears.

Tropical cyclones typically die when they move over cooler water (lack of heat), over land (lack of moisture), or into a region of strong vertical wind shear, such as the midlatitudes.

In addition to damage caused by their extremely high winds, rainfall, and sometimes tornadoes, tropical cyclones can cause landslides, as well as flooding and destruction by the **storm surge**, where sea level rises near the storm due to low pressure and strong wind piling water onto the coastline.

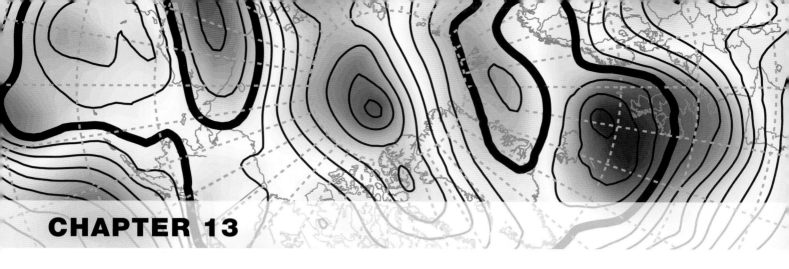

CHAPTER 13

Weather Forecasting

How is it possible to predict, often with great accuracy, upcoming weather? Is a 5-day forecast as good as a 2-day forecast? Can we predict the weather out to the distant future? How is a forecast made? Here we will explore the basic principles behind numerical weather forecasting, as well as the limits of numerical models in accurately predicting weather beyond a certain time. We will discuss the capabilities and limitations of computer models of the atmosphere, methods to address uncertainty, tools to make weather model forecasts more useful, and the role of human forecaster in analyzing and interpreting computer-generated weather predictions.

CONTENTS

One of the things that makes science a uniquely human endeavor is its predictive power. Humans have designed elaborate methods for predicting such things as the motion of planets, tides, and the propagation of diseases, using physics and mathematics. Yet, while we do not question our ability to predict, for example, tides, even though those forecasts have errors in timing and amplitude, we sometimes think of weather as "unpredictable," and we tend to remember poor forecasts even though these are relatively rare. In fact, a close look and a more rigorous assessment of weather forecast accuracy shows that weather forecasters are actually correct much more often than they are wrong, and have been steadily getting better

over the past 50 years. Turning weather prediction from a qualitative art to a quantitative science is one of the major human accomplishments of the twentieth century. In this chapter we explore the basis for the quantitative approach to weather forecasting, which uses the knowledge we have accumulated in the past 12 chapters. This enterprise involves the physical conservation laws we have discussed, a method to solve these equations into the future, powerful computers, and, of course, observations of the atmosphere.

13.1 Weather Forecasts and Uncertainty

We use weather forecasts to make decisions every day. Often those decisions are just about what clothes to wear and whether or not to bring an umbrella. Although the information in those forecasts is mostly deterministic (a single number, such as for the expected afternoon high temperature), this is beginning to change, and there are some aspects that are mostly communicated in the form of probabilities. For example, the probability of precipitation, or POP, might be 30%. This says that there is a 30% chance of measurable precipitation (usually defined as 1 mm or 0.01") at a specific location during a specific period of time. It does *not* say that 30% of the geographical area will receive precipitation, which is a very common misperception. If it rains, does that mean 30% was a bad forecast? What if it does not rain? The way we know if it is a good forecast is by looking at the statistics of many such forecasts and verifying the outcome. We say such a forecast is reliable if, for all of the 30% POP forecasts, measurable rain occurred in 30% of the forecasts. Similarly, a POP of 90% is *reliable* if, for all past 90% POP forecasts, measurable rain occurred 90% of the time. Note that, despite the fact that 90% sounds almost like a "sure thing," we should expect it *not* to rain 10% of the time if this is a reliable forecast.

This example illustrates that weather prediction is a fundamentally probabilistic endeavor. However, the decisions we base on that information often are not: we make yes/no decisions. A familiar example to illustrate how we use probabilities to make decisions, which we don't even think about, involves our behavior when driving a car at different times of day. When

driving on a major highway at 3 a.m., we may take few precautions when changing lanes: there are very few cars on the road, and a quick glance in the mirror is sufficient to confirm that it is safe to move. In contrast, during the evening rush, when the roads are full, we evaluate changing lanes much more carefully. There is a higher probability of something going wrong when the road is full of cars. Here we are also using a rule for our risk tolerance – consequences in an accident are high, so when probabilities for mistakes increase, even a little, our decisions reflect that we are more cautious.

A weather-specific example involves a transportation manager who faces a snowfall forecast and has to schedule overtime pay (expensive) for drivers to treat the highways. If she does not schedule the drivers and heavy snow indeed falls, then there is a high price to pay in the form of a disruption to transit and commerce, which reflects poorly on her job performance. In contrast, if she schedules drivers and it does not snow, there is a price to pay in the form of scheduled, but idle, workers. The snow accumulation forecast can be highly uncertain, say anywhere from 2 to 20 cm, but she still has to decide, yes or no, whether to schedule the drivers, despite the uncertainty. Suppose that she knows that, in addition to the wide range of possible snow accumulation, there is a 30% chance of less than 2 cm, but a 50% chance of more than 10 cm snow accumulation. She can then make a decision based on a risk analysis of the likely cost of both options; in other words, quantifying the uncertainty with probabilities makes her decision easier, and, like the POP example earlier, she will save money over many cases by making decisions based on reliable probabilities.

So, uncertainty and probabilities are fundamental to weather forecasts. Before we explain how we evaluate that uncertainty, we discuss the way individual forecasts are made, which will expose the sources of uncertainty and the methods that have been developed to deal with them.

13.2 Prognostic Equations

Although our daily weather seems complex and ever changing, the atmosphere is a fluid that obeys the laws of physics, in particular the laws of motion, as enunciated by Isaac Newton, the laws of thermodynamics, and mass conservation, as we have established throughout this book. These laws can be cast in mathematical form and used to calculate the future state of the fluid body, given its current state. This can be illustrated using Newton's second law of motion, starting with the familiar form

$$\mathbf{F} = m\mathbf{a} \qquad (13.1)$$

Where $\mathbf{F}$ is the sum of all forces applied to an air parcel (bold indicates vectors), $\mathbf{a}$ is the acceleration (change in velocity) experienced by the parcel, and m is its mass. We can rewrite it as

$$\mathbf{a} = \frac{\mathbf{F}}{m} \qquad (13.2)$$

Now recall that an acceleration is a change in velocity $\Delta \mathbf{v}$ occurring over an interval of time Δt:

$$\mathbf{a} = \frac{\Delta \mathbf{v}}{\Delta t} = \frac{\mathbf{F}}{m} \qquad (13.3)$$

Thus, we can also write that the change in velocity

$$\Delta \mathbf{v} = \frac{\Delta t}{m} \mathbf{F} \qquad (13.4)$$

Since the change is equal to the difference between the future and the current state,

$$\Delta \mathbf{v} = \mathbf{v}_{future} - \mathbf{v}_{current} \qquad (13.5)$$

we obtain that

$$\mathbf{v}_{future} = \mathbf{v}_{current} + \frac{\Delta t}{m} \mathbf{F} \qquad (13.6)$$

In other words, if we know all that is on the right side (the current velocity of the air parcel, its mass, the sum of all forces applied to it, and an interval of time of interest, Δt), then we know the future velocity of the air parcel. We can repeat a similar operation with the other conservation laws.

Thus, if we divide the atmosphere into individual air parcels and choose a time interval Δt of 10 minutes, for example, we could, in theory, calculate the future properties of all the air parcels 10 minutes from now. And if we repeated the calculation by 10 minute intervals, we could extend our forecast to 1 hour, 6 hours, or any number of days. In order to make this calculation, or to use other approaches that

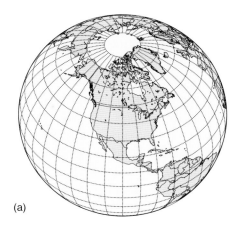

(a)

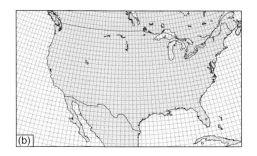

(b)

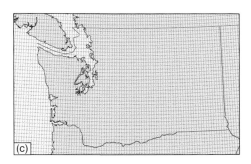

(c)

Figure 13.1. Three examples of grids at increasingly higher resolution. (a) The whole globe covered with a 10° grid. (b) The USA with a 1° grid. (c) Washington State with a 0.1° grid.

do not involve following air parcels, we evaluate the evolution of individual variables (such as velocity, temperature, humidity) at points on a grid that covers a region of interest (Figure 13.1). Coding these equations into a computer at each point of our grid and letting the application run provides a **deterministic forecast**: a single forecast, using a single estimate of the current state of the atmosphere. As we will see, we cannot stop there, because we have not yet measured

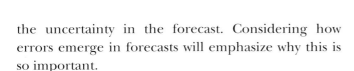

the uncertainty in the forecast. Considering how errors emerge in forecasts will emphasize why this is so important.

Why, then, is the weather forecast for tomorrow sometimes wrong, and the forecast for a month from now completely unreliable? In part, because we do not know *exactly* all that is on the right side of equation (13.6): the **initial conditions**. We either do not know it with enough precision (e.g., instrumental errors) or we do not know it at all in certain locations (e.g., over the oceans, where we have few direct observations). Thus, a small imprecision in the quantities on the right side translates into a small error in the calculation of the future velocity on the left side. Basically, these small errors grow in time to become large errors. Even if we measure the atmosphere thoroughly and accurately, small errors will always be present, grow in time, and degrade the forecast. The limit of forecast skill is thought to be about two weeks, even if we knew the initial state almost perfectly – although there are situations where that skill can be extended to much longer timescales, such as global warming forecasts (see Section 13.4).

In addition to uncertainty in the initial state of the atmosphere, the model representation of the atmosphere is also imperfect. One problem is that we cannot represent all scales of motion exactly; the finest grid cells in Figure 13.1 have a finite size. A second problem is that certain processes, like cloud microphysics, occur at the molecular scale and cannot be resolved explicitly. Electromagnetic radiation is important to the evolution of temperature, but we cannot simulate individual photons. Our grid does not resolve small-scale eddies that contribute to turbulent mixing in the boundary layer. These and other unresolved processes are **parameterized** in terms of the scales of motion we can resolve with the chosen grid. This approximation translates into a significant level of uncertainty in the model. These model errors come in addition to errors in the initial conditions. In summary, we can see that weather models are complex computer algorithms subject to uncertainty in their construction, in addition to the uncertainty in the initial data for the forecast.

The amplification of errors in the characterization of the atmosphere is illustrated in Figure 13.2, where

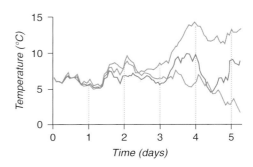

Figure 13.2. Three temperature forecasts with slightly different initial conditions.

three different forecasts are made with slightly different initial conditions and the temperature at some weather station is plotted against time. In this example, the predicted temperatures are almost identical on the first day, and relatively similar on the second day, although they start to depart from each other at the end of the day. By the end of day 3, the difference is as large as 3 °C, while beyond the fourth day the temperatures span a range of more than 10 °C. Therefore, in this case we can have **confidence** in our 1-day forecast, and even our 2-day and 3-day forecasts to some extent, because in spite of the errors that might be present in our initial data, the actual weather will most likely evolve within the envelope of the three curves and the actual temperature will fall within a 3 °C window. Beyond 3 days, however, the actual temperature might be anywhere within the 10 °C window suggested by the different forecasts, and our confidence is greatly reduced.

This decreasing accuracy of weather forecast models beyond a few days is illustrated in a different way in Figure 13.3. Rather than positioning ourselves at time 0 and watching a forecast evolve into the future, we now choose a time of interest (here, our midlatitude cyclone on February 21, 2014, at 00:00 UTC), and look back at how *different* forecasts performed, starting from different points in time. Since the actual weather event has occurred, we have the observed pressure and temperature field (Figure 13.3(d)) against which we can verify the different forecasts.

Figure 13.3(a) shows a 5-day forecast computed on February 16, 2014, at 00:00 UTC. Obviously, the forecast did not perform very well, since the

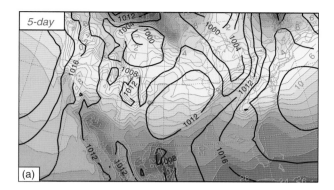

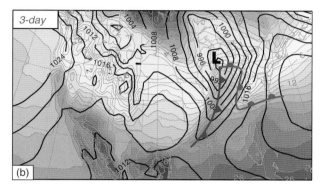

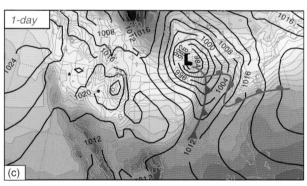

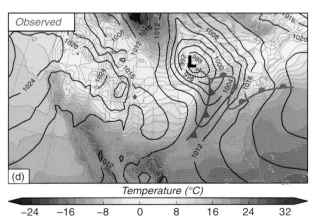

Temperature (°C)

−24   −16   −8   0   8   16   24   32

Figure 13.3. (a)–(c) Comparison of different forecast runs for the same valid time, February 21, 2014, at 00:00 UTC. (d) The observed pressure and temperature fields on February 21, 2014, at 00:00 UTC.

cyclone is completely missing, except for some sign of a low pressure trough extending south between two highs.

The 3-day forecast shown in Figure 13.3(b) is significantly improved. Our mature cyclone is present with fully developed low and fronts. Yet, the warm front is positioned too far north, and the cyclone is not showing any sign of occlusion. In other words, the predicted storm development is too slow: the actual cyclone will develop and occlude faster than predicted.

Not surprisingly, the 1-day forecast is much more accurate (Figure 13.3(c)), and the differences with the control analysis are in the fine details. Indeed, current models perform extremely well within 24 to 48 hours, and certain large-scale aspects of the weather are predicted well beyond a week, as we will see.

Now that we have the basis for making a quantitative weather forecast, and the sources of error and uncertainty in that forecast, we will return to how to quantify that uncertainty and provide the probabilities we need to make decisions.

13.3 Ensemble Forecasting

As we have seen, if we run a number of weather forecasts with slightly different initial conditions that reflect our uncertainty about the initial data, we can compare all the forecasts and gain confidence where they mostly agree, while exercising caution where they mostly disagree. Similarly, we can adjust the parameterizations of the unresolved processes in the model, and make new forecasts to assess how that source of uncertainty affects the forecast. The group of forecasts that results from this suite of computations is called an **ensemble**, and each individual forecast is called a **member** of the ensemble. Ensemble forecasting is the modern method by which weather forecasts are made and probabilities are estimated.

To continue with the previous example in Figure 13.2, where we ran an ensemble forecast with three members, we could issue a forecast of 6 °C after day 1 with confidence, 7–8°C after day 2, 5–8°C after day 3, while we would suggest 6–13°C and 3–13°C for days 4 and 5, respectively. A more mathematical way of doing this would be to calculate the statistical **uncertainty** associated with the forecast. Here, we cannot do

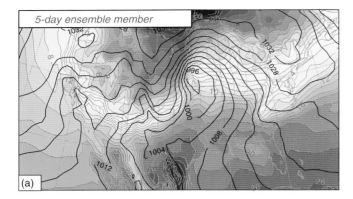

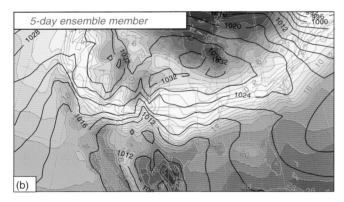

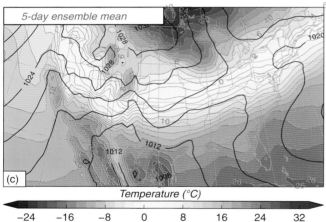

Temperature (°C)

−24   −16   −8   0   8   16   24   32

Figure 13.4. (a), (b) Two examples of particularly different 5-day forecast ensemble members for the same valid time as in Figure 13.3, February 21, 2014, at 00:00 UTC. (c) The 5-day ensemble mean for the same forecast time.

statistics with three members, but, for the sake of illustration, the 1-day forecast might be 6±1 °C (i.e., 6 °C plus or minus 1 °C), while the 4-day forecast might be 8±5°C (i.e., a greater uncertainty).

Returning to our 21 February 2014 cyclone, Figure 13.4 shows that the uncertainty in forecasts of

this event is unusually large. The two outlier ensemble members are chosen to illustrate why looking at a single model forecast can be very misleading (Figures 13.4(a) and (b)). One ensemble member has a large and intense midlatitude cyclone, whereas the other has a large and strong anticyclone. A forecaster relying solely on one of these forecasts would make drastically different forecasts just 5 days into the future. Figure 13.4(c) shows the average over all of the ensemble members, which is the best forecast we have because it averages over all of the random differences in the ensemble members and leaves the main signal that they have in common. In this case, the uncertainty is unusually large and the ensemble mean shows a large area of high pressure across the region where a major cyclone would occur in just 5 days. Some of the ensemble members had the right idea of a major cyclone developing, but they were in the minority, and there is no way to know in advance which ensemble members are the ones to base a forecast upon.

Another way to represent the forecast uncertainty from an ensemble is to plot a particular contour for all ensemble members, which results in what are called "spaghetti" plots. This is common on upper-level maps, and an example for a single 500 hPa contour for our cyclone case is shown in Figure 13.5. In this particular example, the 5240 m contour (at 500 hPa) is shown in black for different forecast ensemble members with slightly different initial conditions. We can see that the spread is relatively small in the 1-day forecast (Figure 13.5(a)), but increases as we go to 3 days (Figure 13.5(b)) and 5 days (Figure 13.5(c)). In the 5-day forecast we notice that all ensemble members agree relatively well in places, like the North Pacific Ocean (narrow spread), but disagree in other places, like the North Atlantic Ocean and the Norwegian Sea (wide spread). It tells us that we should be cautious about drawing conclusions and predictions that are strongly dependent upon the position of upper-level features in the North Atlantic, such as the position and intensity of a surface low.

One drawback with spaghetti diagrams is that the lines tend to spread out more in regions where the gradient in the field is small (lines spaced far apart) as compared to regions where the gradient is large (lines spaced close together). This simply indicates

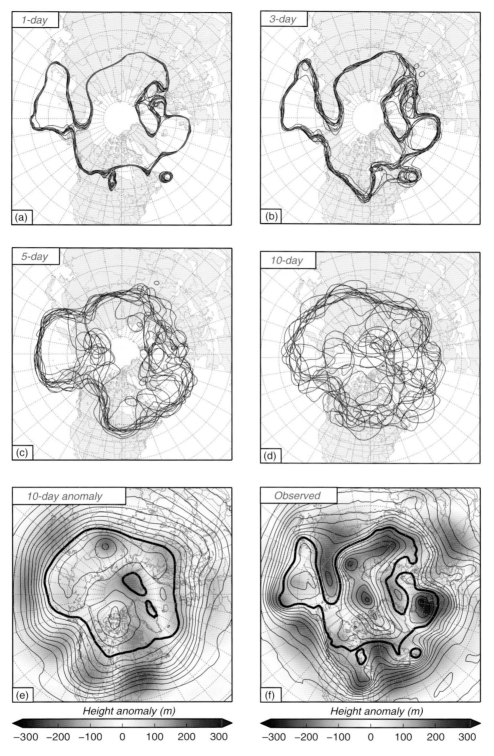

Figure 13.5. Comparison of different ensemble forecasts for the same valid time, February 21, 2014, at 00:00 UTC. (a)–(d) The 5240 m contour of the 500 hPa surface is shown in black for 11 members. The red contour shows the ensemble mean. (e) The 10-day ensemble mean 500 hPa height field. (f) Observed 500 hPa height field on February 21, 2014, at 00:00 UTC. In (e) and (f), the 5240 m contour is highlighted in bold, and colors indicate the height anomaly with respect to the 30 year mean 500 hPa height field for the month of February.

that small changes in the value move the contour a lot, which is not indicative of large uncertainty.

The 10-day forecast seems relatively useless at first glance (Figure 13.5(d)). The 10-day ensemble mean, however, contains some useful information at larger scales. This is shown in Figure 13.5(e), where the 10-day ensemble mean 500 hPa height field is overlaid on the "anomaly" with respect to the climatological mean for February: that is, the difference between the ensemble mean and the climatological mean. Blue areas indicate lower heights than average for February, and red areas indicate higher heights than average for February. We can see that the 10-day ensemble mean indeed contains some indication of an upper-level trough over the western half of the USA (blue in Figure 13.5(e)), and of an upper-level ridge over New England (red in Figure 13.5(e)). This pattern roughly corresponds to the observed synoptic situation and anomalies on February 21, 2014, at 00:00 UTC (Figure 13.5(f)), so the 10-day forecast contains useful information about general changes in the pattern, although it lacks crucial details about individual weather features, such as our midlatitude cyclone.

In conclusion, even though we lose specific details about individual weather features for forecasts out to 10 days, there is still useful information about large-scale patterns that we can glean from studying the ensemble members, the ensemble mean, and the anomaly of the mean with respect to the climatology.

13.4 Chaos and Weather Prediction

Will we ever be able to make a perfect deterministic weather forecast if, one day, we can observe the atmosphere in its smallest details and build a computer that can resolve the smallest scales? Probably not. It is generally accepted that it is technically impossible to predict details of the weather beyond about two weeks. And that is not due to human limitations in observing or modeling the atmosphere; it is due to the nature of the atmosphere itself. In the end, the amplification of initial errors is due to the *chaotic* nature of the atmosphere, which is described by **chaos theory**. Note that, in this context, chaos does not mean "total disorder,"

as we often imply in daily parlance. Rather, chaos means that the prediction of the state of the atmosphere at future times depends sensitively on the details of the atmosphere at the starting time of the forecast, as we saw quite dramatically in Figures 13.4 and 13.5. In other words, our inability to eliminate *all* error from the initial state inevitably leads to the growth of those errors that ultimately degrade the forecast.

For the atmosphere, the largest scales of motion, for example planetary-scale waves on the jet stream (Figure 13.6(a)), have the longest predictive timescale, and the smallest scales of motion (Figures 13.6(e) and (f)) have the shortest predictive timescale. This is due partly to the fact that small spatial scales naturally have short timescales. For example, gusts of wind may last a few seconds whereas changes in the wind in planetary waves may take place over many days. A second fact is that the largest scales have much more energy than the small scales; the total energy in all of the small wind gusts is dwarfed by the energy in the jet stream. As a result, errors of a given size take much longer to "saturate" at the larger energy values found at planetary scales compared to small scales. Therefore, while small-scale errors would eventually grow and become significant at larger scales in a longer-term forecast, they remain relatively small compared to larger scales in a 3 to 5-day forecast.

Although we have painted a gloomy picture of improving weather forecasts beyond about two weeks, there is an important exception to this rule. If something occurs that "forces" the atmosphere, then predictive skill may result if the forcing is known or may be predicted. This forcing is usually something that occurs on longer timescales than those that dominate weather variability in the atmosphere. Two examples are global warming and the seasonal prediction of the El Niño/Southern Oscillation (ENSO) phenomenon.

For global warming, all forecasts of specific weather events (like a particular cyclone) are still subject to the two-week limit, but the warming itself, which is essentially a known forcing from greenhouse gases, can be well forecast out to decades and centuries on the global scale (see Chapter 15). This leads to changes in the jet stream and, consequently, in the location of storm tracks and weather events. Again, we cannot predict specific weather events, but these global

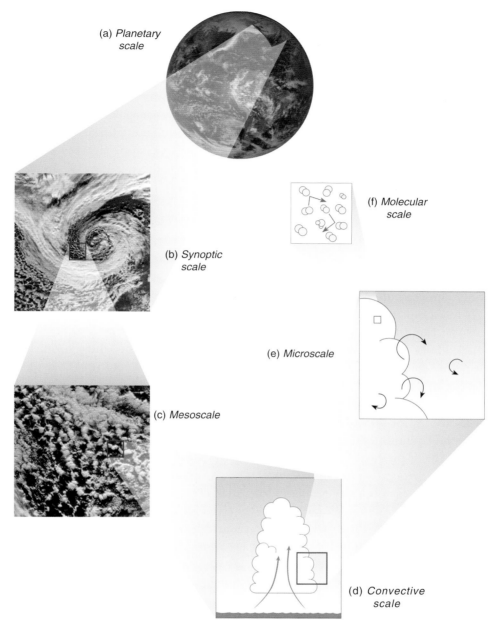

(a) *Planetary scale*

(b) *Synoptic scale*

(c) *Mesoscale*

(d) *Convective scale*

(e) *Microscale*

(f) *Molecular scale*

Figure 13.6. Atmospheric motion organized at various scales: small spatial scales have shorter time scales and less energy than large scales of motion.

warming forecasts can inform us about the change in the statistics of weather events. For example, if the jet stream shifts poleward, then regions that will experience the effects of extratropical cyclones will also shift poleward.

For ENSO, changes in the temperature of the tropical Pacific upper ocean force changes in the atmosphere through teleconnections related to planetary waves. These planetary waves are excited by convection that occurs over unusually warm water. Since the evolution of the upper ocean temperatures is a process taking place over 6–12 months, and since it often does so in a predictable way, it provides some predictive skill to weather averages on these timescales. Again, we cannot predict individual weather events, but the statistics of weather features, such as temperatures

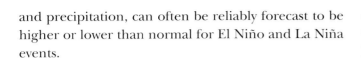

and precipitation, can often be reliably forecast to be higher or lower than normal for El Niño and La Niña events.

13.5 From Forecast Grids to Reliable Forecast Values

Once the computation for an ensemble forecast has been completed, the raw result is a grid of numbers for temperature, pressure, wind, precipitation, etc. While we could use those numbers for making a weather forecast, it turns out that another important step is usually taken: **statistical post-processing**. The "post" part of the processing reflects the fact that this step takes place after the computer simulations have been performed. It turns out that some of the errors in the forecast are not random, but systematic, and we can increase the skill of our forecasts if we remove the systematic error. One source of systematic error is called **bias**. We often find that weather forecasts from the computer models are systematically too warm or too cold for some locations; for others, there may be too much or too little precipitation. Often these biases result from a complex set of processes in the models that are very difficult to correct. For example, errors in the treatment of clouds can produce these biases in temperature and precipitation, but since the physics of clouds depends on many parts of the model, including those that are parameterized (e.g., cloud microphysics and radiation), it is very difficult to fix these biases without making other aspects of the forecast worse.

Bias may not be obvious in any individual forecast because other sources of error may be much larger, but it becomes apparent when we average the errors over many forecasts. By estimating bias in this way, and removing it from the raw computer forecasts, we can produce forecasts that are more accurate. This is an example of **forecast calibration**. Another example of forecast calibration involves the uncertainty estimate derived from the ensemble forecast. We often find that the predicted uncertainty from the range of forecasts in the ensemble is too small. This is due to the fact that some errors are deterministic, which reduces the range of uncertainty that the ensemble should capture. For example, if there are no errors coming from the land surface below the atmosphere, that tends to reduce ensemble variability in near-surface fields. Again, as with bias, if the ensemble forecasts are verified over a large number of forecasts, that systematic error can be corrected, resulting in calibrated ensembles.

Another very successful approach in statistical post-processing is called **model output statistics** (MOS). Rather than making forecasts directly from the model grids, the MOS approach involves using a set of predictors from these grids to estimate the quantities of interest. For example, we want daily high and low temperature forecasts at specific locations, and we could just use the raw model output, or its bias-corrected version, directly. Instead, MOS uses a wide range of quantities that are known to be important and directly related to high and low temperature forecasts. For example, temperatures, winds, and clouds throughout the lower atmosphere are known to be important predictors of surface temperature. Using a large sample of historical forecasts of these predictors, and the actual high and low temperatures that were observed, MOS builds a calibrated statistical relationship that allows us to predict the high and low temperature better. These relationships are custom to each location and season, which insures they are optimal for each location. One reason this method works is that it provides an efficient way to remove bias from a large number of fields that may contribute to temperature forecasts, but it also accounts for the fact that the points for which we want a forecast are rarely at model grid points.

MOS provides an important contribution to actual weather forecasts, and it is difficult for humans to intervene and improve upon these forecasts. One situation where humans can, in fact, beat MOS forecasts is during extreme events. When the forecast extends beyond where MOS has been calibrated, it can lead to unusually large errors. For example, record high temperature forecasts mean that the value has not been observed previously. It is in these situations, which often have a high impact on society, that the human forecaster plays an outsized role in the forecast. Another place where humans play an important role is in very short-range forecasts, which we call **nowcasts**. If observations reveal that the most recent computer forecasts are in error, the human forecaster can correct for that. Newer techniques, based on **machine learning**, are also under development and promise to improve further our ability to post-process weather forecast information.

Once we have completed statistical post-processing of the forecast information, it is time to make the forecast. As we have indicated, it is difficult for the human forecaster to improve upon this forecast information, with important exceptions. Instead, the main role of the human forecaster is to understand why the forecast is as it is, and to explain that to society. To do so, human forecasters often follow a method of building a conceptual understanding of the forecast, and we now discuss a common method that you can use as well.

13.6 Making a Forecast

To conclude this chapter, let us survey some of the tools meteorologists use to issue their forecasts. As a general rule, when we prepare a weather forecast, we usually start with the largest scales and then zoom in on a particular region of interest. Thus, it is common to obtain first a view of the entire hemispheric circulation, commonly using the **500 hPa height field**, for example, as shown in Figure 13.7(a). It is helpful to look at a **loop** going back a few days in the past, to follow where the troughs and ridges are coming from, and extending a few days into the future, to see where these features are going. This overview helps us to appreciate the larger-scale context in which our local weather is occurring, and becomes increasingly important as we consider forecasts further in the future.

Satellite imagery is a good complementary resource, as it provides us with a hemispheric view that we can use to assess quickly the amount of cloud and water vapor associated with the 500 hPa troughs that we have previously identified. Midlatitude cyclones at different stages of their life cycle are evident, because we know their cloud signature (Figure 13.7(b)). By comparing **visible** and **infrared** imagery, we can identify fog, low stratus, and cumulus versus high cirrus, as well as thunderstorms.

With this broad picture in mind, we can now zoom in on a particular region of interest and study the **surface pressure** and **temperature** fields in more detail. We often combine them in different ways to obtain a synthetic view. For example, combining 500 hPa heights with surface pressure facilitates comparison between the position of upper-level troughs and surface lows (Figure 13.7(c)). At this point we will focus

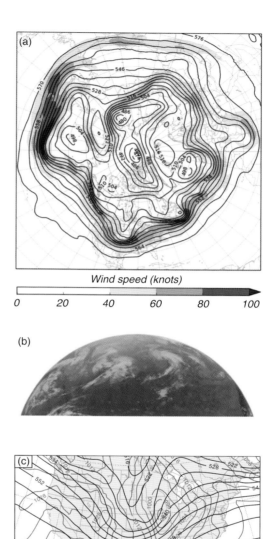

Figure 13.7. Some of the maps and images a weather forecaster might analyze in time-lapse animations. (a) Hemispheric map of 500 hPa heights. (b) Infrared satellite imagery. (c) 500 hPa heights and mean sea level pressure.

on specific features that we know to be locally important. For example, if a front is moving through during the forecast period, we will look at **weather station data** to characterize the strength of the front in terms of temperature contrast and wind, and the type and intensity of the precipitation field.

If there is a chance of snow in the forecast, we will want to know the exact position of the low, because that

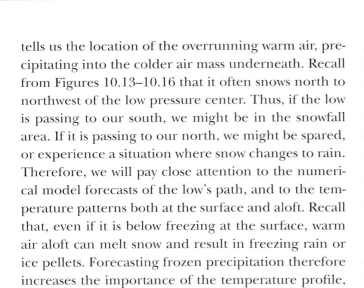

tells us the location of the overrunning warm air, precipitating into the colder air mass underneath. Recall from Figures 10.13–10.16 that it often snows north to northwest of the low pressure center. Thus, if the low is passing to our south, we might be in the snowfall area. If it is passing to our north, we might be spared, or experience a situation where snow changes to rain. Therefore, we will pay close attention to the numerical model forecasts of the low's path, and to the temperature patterns both at the surface and aloft. Recall that, even if it is below freezing at the surface, warm air aloft can melt snow and result in freezing rain or ice pellets. Forecasting frozen precipitation therefore increases the importance of the temperature profile, which we monitor closely through **radiosonde profiles** and other measurements.

Temperature profiles are also useful when forecasting the probability of fog and how long it will last before burning off. If there is a temperature inversion and the dew point is high with little wind, fog is more likely to remain trapped. This is especially true when the sun is low in the sky, such as during winter. Similarly, temperature profiles are critically important when forecasting the probability of thunderstorms, as they tell us about the stability of the atmosphere and how much buoyancy is available for the development of deep cumulonimbus clouds. So we look not only at radiosonde profiles, but also at temperature profiles produced by numerical model forecasts, which can provide a guide to the timing and location for convection to develop. This problem remains, however, one of the most difficult high-impact weather events to forecast accurately.

When precipitation is forecast, **radar images** are very instructive, as they tell us about the exact location of the rain patterns, their width and structure, and the speed at which they are moving. Specific signatures in the radar echoes provide information that can indicate the presence of hail, or the level in the atmosphere where snow is melting.

Once the forecaster has developed a good familiarity with the features that will affect the forecast, including the observed history and projections into the future from numerical model forecasts, it is time to produce a specific forecast. Common elements of a forecast consist of quantitative predictions for low temperature, high temperature, and probability of precipitation.

Additional factors such as wind speed and cloud cover are often also provided. If MOS is available for the forecast location, this is typically the best starting point for such forecasts, as it is hard to beat on average. As we discussed, though, there are situations where the forecaster can and should adjust MOS predictions, such as when observations show the model forecast is off track, during extremes, or in certain situations where MOS itself is known to have a bias. (MOS only removes longer-term bias, not bias conditional on specific weather features.)

Perhaps the most important thing that a forecaster provides, and this cannot be automated, is a discussion of the forecast that provides the context and physical basis for the forecast, an assessment of forecast uncertainty (or confidence in the forecast), and an indication of what can go wrong. This is often done by broadcast meteorologists, as well as being routinely found in the **Area Forecast Discussion** issued by local offices of the US National Weather Service. A frequent, high-impact, example is winter snowstorms, where small changes in the track of the low center, for example, can have a large impact on snow-amount forecasts at particular locations. In that case, as discussed earlier, the forecaster should review the information coming from ensemble forecasts, which sometimes shows that the uncertainty is skewed in a particular direction. For example, although there may be uncertainty in the track of the low center, the ensemble may show a preference for a southern track as compared to a northern track, and that is useful information to communicate in the forecast.

13.6.1 Medium to Long-Range Forecasting

Even though it becomes difficult to forecast the details of upcoming weather beyond about 5 days, weather services often issue a longer-term forecast called the **medium-range forecast**, typically one to two weeks in advance, on more general aspects, rather than the details found in a short-term forecast. Instead of trying to predict exact temperatures, precipitation amounts, or snow depth, medium-range forecasters look for critical features that are persistent in the model forecasts, from run to run, and from model to model. As such, ensemble forecasts are the primary tool for medium-range forecasts.

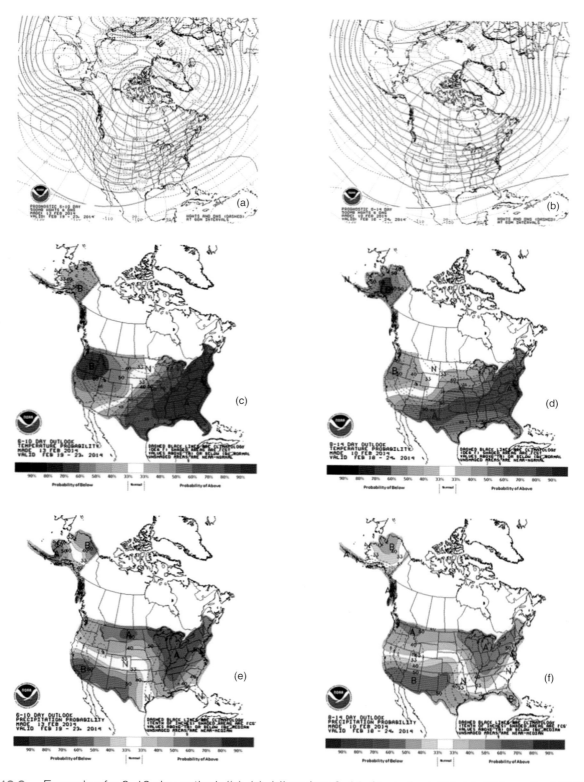

Figure 13.8. Example of a 6–10-day outlook ((a), (c), (e)) and an 8–14-day outlook ((b), (d), (f)) from the NOAA Climate Prediction Center corresponding to the four-day average conditions centered on February 21, 2014, at 00:00 UTC. (a), (b) 500 hPa heights (solid green contours) and anomalies (dashed red contours). (c), (d) Temperature probability. (e), (f) Precipitation probability.

These forecasts usually associate large-scale patterns with changes in climatological (historical average) probabilities for what normally happens at the particular time of year. A persistent upper-level ridge, for example, is often indicative of a greater chance of clear skies, and weaker winds. The presence of an upper-level trough, however, may suggest a greater chance of clouds and precipitation. Forecasters might mention that chances for precipitation are higher or lower, without explicitly detailing the features that will produce the precipitation. Figure 13.8 shows such an example of a 6–10-day outlook and an 8–14-day outlook issued by the NOAA Climate Prediction Center, and corresponding to our February cyclone.

These longer-term forecasts are becoming more accurate and valuable as computer power increases and computer models improve our ability to perform ensemble forecasting.

13.6.2 Seasonal Outlook

Similarly, forecasters can use the departures from climatology based on long-term indicators such as teleconnections to issue a **seasonal outlook** several months in advance. The main source of predictive skill at these timescales is associated with ENSO. For example, increasing sea surface temperatures in the Central Pacific Ocean, announcing the return of an El Niño event, can be the basis for predicting higher chances of rain and flooding for California in late winter (i.e., higher chances than average). Here again, forecasters are not trying to forecast the weather on a specific day three months in advance using the techniques described for short-term forecasting. They are mainly using statistical techniques to determine if weather variables will be average (climatology), higher than climatology, or lower than climatology (i.e, the seasonal-average conditions over a long period of time, typically 30 years). These seasonal outlooks can be valuable for farming, energy planning, recreational activities, and risk preparedness, for example. Again, with increases in computer power, numerical models are now being used, coupled to ocean prediction models, to make seasonal forecasts.

Summary

The atmosphere is a fluid that obeys the laws of physics, and the mathematical form of these laws can be encoded for computers to solve, yielding a future prediction given the current state of the atmosphere. This involves calculating the evolution of individual variables (such as velocity, temperature, humidity) at certain points on a grid that covers a region of interest. Weather observations provide the starting point for the forecast.

Our ability to forecast future weather is limited because:

▶ the observations used as **initial conditions** cannot be known exactly;

▶ certain small-scale processes are not resolved by the model grid and must be **parameterized**, which translates into uncertainty and model errors; and

▶ small errors in the initial conditions grow in time to become large errors, ultimately completely contaminating the forecast.

The limit of weather forecasting skill is thought to be about two weeks, even if we know the initial state almost perfectly. **Ensemble forecasts**, where many forecasts are generated from slightly different initial conditions, and by different forecast models, provide the tool by which we measure the loss of skill in the forecast. Typically, small differences among the different forecasts grow quickly, so by 5–10 days we cannot usually speak with confidence on details in the forecast, but often there are broader patterns of agreement for which we do have forecast **confidence**. Persistent features allow weather forecasters to issue **medium-range forecasts** out to two weeks.

In spite of the **chaotic nature** of the atmosphere, forecasting its future state is possible on longer timescales because there exist **forcings** that affect, or constrain, the dynamics of the atmosphere and therefore the statistics of weather events (e.g., global warming and ENSO). These forcings allow forecasters to issue a **seasonal outlook** several months in advance and to make forecasts of climate for several decades. Again, individual weather features cannot be predicted on these timescales, but we can often make useful predictions about how the weather, on average, will depart from historical averages (climatology).

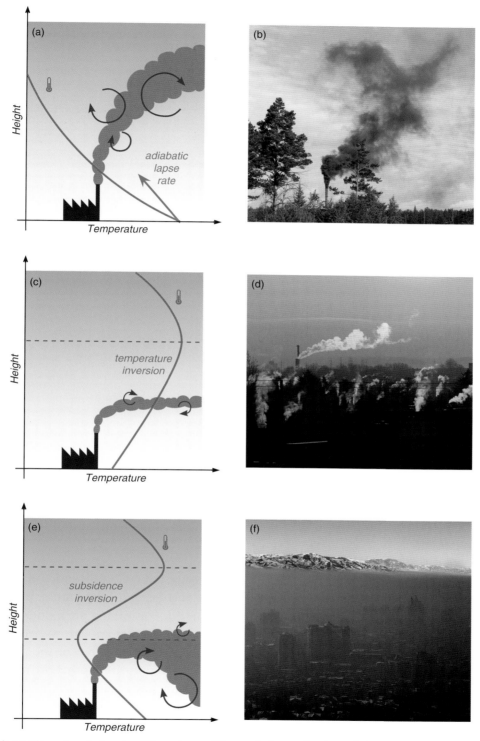

Figure 14.3. Dispersion of pollution in different conditions of atmospheric stability. (a), (b) In unstable, convective conditions, the pollutants are quickly mixed and transported upward. (c), (d) In stable conditions, the pollutants spread out horizontally. (e), (f) If the atmosphere is unstable under a subsidence inversion, the layer of pollution is trapped and mixed downward. (a), (c), (e) Temperature profile shown in red.

14.3 Large-Scale Patterns

Our last example illustrates the role played by large-scale weather patterns in controlling the fate of pollution. Anticyclones are often associated with a stable atmosphere (i.e., no vertical transport) and light to no winds (i.e., no horizontal transport). Therefore, pollutants tend to stagnate and to increase in concentration if anticyclonic weather persists. That is often the case in Los Angeles, California, for example, due to the presence of the Pacific High. If a subsidence inversion forms, with some downward mixing of pollutants as described in Figures 14.3(e)–(f), the layer of pollution can become trapped and extend to the surface. In anticyclonic conditions, the atmosphere is also often free of clouds, which allows sunlight to penetrate and cause photochemical reactions, leading to the formation of photochemical smog. Thus, the combination of all these factors can be conducive to a persistent health hazard, as is well known of cities such as Los Angeles and Mexico City (Figure 14.4).

In contrast, low pressure systems are associated with convergence and vertical motion, as well as stronger pressure gradients and wind. Therefore, pollutants tend to be mixed efficiently and transported upward out of the mixed layer, where they join the jet stream and may be quickly transported away from the source (recall Figure 14.1). Additionally, clouds block sunlight and reduce the potential for photochemical reactions, and precipitation scavenges aerosols. Therefore, stagnating pollution and photochemical smog are typically less of an issue in cyclonic weather conditions.

14.4 Topography

The concentration of pollutants in the lower troposphere is sometimes made worse by topography. In particular, recall that mountain slopes cool by emitting infrared radiation (or by emitting more infrared radiation than they absorb visible light). As the adjacent air layer also cools (by conduction), it becomes relatively denser and slides downward toward the valley bottom (Figure 8.19). This cold air tends to accumulate in the valley, especially on cold, clear nights, and even more

Figure 14.4. Photochemical smog in Los Angeles, California.

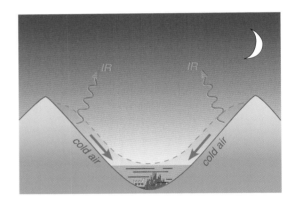

Figure 14.5. Concentration of pollution at the bottom of a valley.

so during anticyclonic conditions. In valleys harboring industries and urban areas, pollutants follow this downward, converging air motion and tend to accumulate at the bottom with the cool air in the valley (Figure 14.5). This often contributes to the formation of photochemical smog in Los Angeles (surrounded by the San Gabriel and Santa Monica Mountains) and Mexico City (surrounded by Las Cruces and Ajusco Mountains and the Sierra Madre), for example.

During the day, the mountain slopes warm by absorption of solar radiation, and the adjacent air, warmed by conduction, moves upslope, draining the pollutants out of the valley, as it were. But this only happens if the warmer air is sufficiently buoyant to break the temperature inversion.

Summary

Some chemicals and aerosols are particularly detrimental to human health and the environment.

▶ **CFCs** destroy the ozone layer in the stratosphere and allow more ultraviolet radiation from the sun to penetrate through the troposphere.

▶ Greenhouse gases such as **carbon dioxide** (CO_2), **methane** (CH_4), and **nitrous oxide** (N_2O) play a major role in global warming.

▶ **Carbon monoxide** (CO) prevents the absorption of oxygen in the blood, which can lead to death.

▶ **Sulfur dioxide** (SO_2) can be oxidized in the atmosphere to form sulfuric acid (H_2SO_3) and acid rain. **Nitric oxide** (NO) and **nitrogen dioxide** (NO_2), more generally referred to together as **NO$_x$**, can also contribute to acid rain.

▶ **Tropospheric ozone** (O_3) in high concentrations causes eye irritation and respiratory disorders, and contributes to *photochemical smog*.

▶ Some **volatile organic compounds** (VOCs) are carcinogenic, while others contribute to photochemical smog.

To this list of gases we add solid particles (**aerosols**) emitted by combustion. **Particulate matter** of diameter less than $10\,\mu m$ (**PM$_{10}$**) concentrates in our lungs and causes a number of respiratory problems. Fine particles of diameter less than $2.5\,\mu m$ (**PM$_{2.5}$**) can pass *through* our lungs. Both are regulated by the US **Environmental Protection Agency**.

Certain pollutants take part in complex chemical reactions in the presence of sunlight to produce a noxious mixture of secondary pollutants called **photochemical smog**, containing ozone, NO$_x$, and VOCs, among others.

Gases and particulate matter are transported and dispersed horizontally by the local wind circulation, or vertically by convection and upward motion in mid-latitude cyclones, up to the jet stream, and around the hemisphere.

Several factors affect the dispersion of pollutants:

▶ The rate and scale of mixing vary with the **stability** of the atmosphere.

▶ When the atmosphere is **unstable**, exhaust plumes rise and mix with clear air, and diffuse higher in the atmosphere.

▶ In the presence of a **temperature inversion**, exhaust plumes tend to flatten out horizontally in the inversion layer.

▶ When the inversion layer caps an **unstable** layer underneath, the pollution is mixed downward by turbulent motion, which distributes the pollutants over the whole depth of the lower layer (i.e., the **mixed layer**).

In anticyclonic weather (i.e., stable conditions), pollutants tend to stagnate and to increase in concentration, and the formation of photochemical smog is common in cities and industrial areas. In cyclonic weather (i.e., strong winds and vertical motion), pollutants tend to be mixed more efficiently.

Pollution tends to be drained down and accumulate into valleys by the sinking of cold, dense air along the slopes at night, and to be more mixed, and thus diluted, during the day.

CHAPTER 15

Climate Change and Weather

How will weather change as our planet warms and our climate changes? Will there be more droughts? How will precipitation patterns be affected? Will there be more or fewer storms? Will tropical cyclones be more, or less, intense? Will all regions of the world be affected? In this last chapter, we will put in perspective all that we have learned about the atmosphere and apply it to the pressing issue of global warming.

CONTENTS

As we close the exploration of our atmosphere and weather, we must pause and ask ourselves how they will evolve in the near future. Will weather be any different on a warmer planet?

Throughout this book, we have come to understand that weather results largely from temperature contrasts and energy transfer. Thus, since weather is ultimately about energy redistribution, and global warming involves an increase in the energy of the atmosphere, it is legitimate to wonder how this additional energy will affect existing weather patterns. Although it is beyond the scope of this book to provide a complete description of global warming and its impacts, we will use this topic to apply what we have learned to understand current thinking about the most likely evolution of our weather in a warmer world.

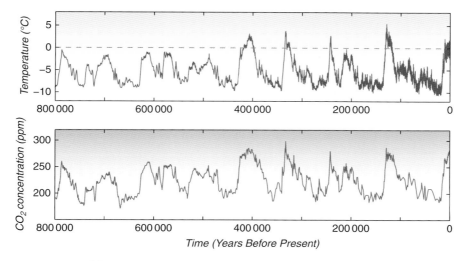

Figure 15.1. Temperature and CO_2 variations over the last 800 000 years reconstructed from the analysis of an Antarctic ice core.

15.1 Past and Future

Earth's climate has changed in the past and will continue to change in the future. Although the time period over which humans have measured climate directly with scientific instruments is limited to about the last 150 years, the history of Earth's climate is recorded indirectly, or by proxy, by living organisms and geological processes. For example, analysis of the isotopic composition of water molecules in Antarctic ice cores reveals that Earth's climate has gone through many cold and warm periods over at least the past 800 000 years (Figure 15.1). We know that these temperature variations are due in part to cyclical changes in Earth's orbit around the sun, as well as its own rotational axis, producing changes that take place over thousands of years. Earth's path on the ecliptic plane, for example, does not always have the same shape. It is at times more elongated (i.e., more elliptical), and at other times more circular, following a cycle of about 100 000 years. As a result, Earth is at times closer and at other times farther away from the sun and receives more or less solar energy in different seasons. The tilt of Earth's axis with respect to the ecliptic also changes, with the North Pole pointing at times more directly toward the sun, resulting in the summer hemisphere receiving more sunlight, and at times less, with a cycle

of about 40 000 years. And Earth's position and orientation after one revolution around the sun (i.e., after a year) also change over time with a cycle of about 26 000 years, resulting in the seasons starting at different points of the ellipse throughout the cycle. Because all these cycles happen over different periods of time, they combine in a complex way to produce a varying input of energy into the Earth system and, importantly, varying seasonal temperature contrasts between the two hemispheres.

Even though they are complex, however, orbital changes happen with cyclical regularity and can be computed accurately (something the Serbian mathematician Milutin Milankovitch did as early as 1920). It turns out that orbital changes alone are not sufficient to explain the observed temperature variations of the last 800 000 years. To explain the 8–10°C temperature variations observed in Figure 15.1, an additional process needs to take place that enhances the temperature increase or decrease associated with orbital changes, i.e., a positive feedback mechanism.

We now know that this positive feedback is in part a greenhouse effect feedback due to the release and intake of CO_2 initiated by the temperature variations. You can verify in Figure 15.1 that the CO_2 concentration indeed increases when the temperature increases, and decreases when the temperature

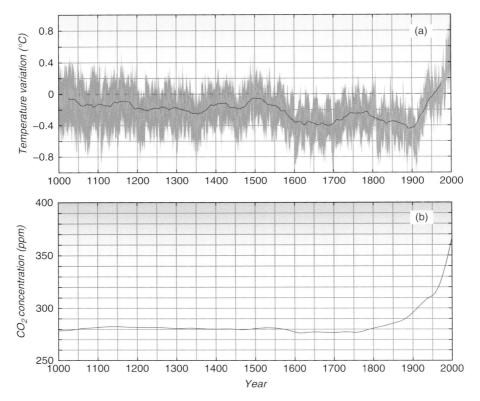

Figure 15.2. (a) Temperature variations (brown), 50 year running average (red), 5th and 95th percentile (gray) over the past 1000 years. (b) Variations in CO_2 concentration over the same period.

decreases. Once the positive feedback loop is initiated, of course, each effect reinforces the other, as we described in Box 4.5.

Thus, the CO_2 concentration responds to changes in temperature, and both can vary in response to a number of natural phenomena, such as orbital changes and volcanic eruptions: that is what we call the **natural variability** of the climate system. But temperature can also respond to changes in CO_2 concentration, and, most relevant for global warming, to the anthropogenic changes that began with the onset of the Industrial Revolution (Figure 15.2). To understand better this **forced variability** of the climate system, we develop numerical models of the climate system to gain insights into future climate changes resulting from increased greenhouse gas concentrations and global warming.

We will now explore these predicted changes in relation to the topics we have explored throughout this book.

15.2 Changing Composition

We recall from Chapter 3 that the atmosphere is composed of permanent gases (nitrogen, oxygen, and a number of trace gases) and variable gases such as water vapor, carbon dioxide, and ozone (among others). Some of these variable gases contribute to the greenhouse effect, such as water vapor, carbon dioxide, methane, nitrous oxide, tropospheric ozone, and CFCs. Many of these variable gases have increased in concentration over the last 200 years, in varying amounts. While the CO_2 concentration, for example, has increased by 40% compared to pre-industrial levels, from about 280 to 400 ppm, the concentration of methane has increased by 150% from about 700 to 1800 ppb (parts per billion).

If we place 400 ppm in Figure 15.1, for example, it is clear we are now seeing CO_2 concentrations that

Earth has not experienced in the last 800 000 years. We are heading toward a future for which we cannot find an analog in the recent climate archive.

15.3 A Warmer World

We saw in Chapter 4 that an increase in greenhouse gas concentration results in an increase in the heat content of the atmosphere through the enhancement of the greenhouse effect. The global average temperature has already increased by about 1 °C over the past 150 years (Figure 15.2), and current climate models suggest that, depending on what decisions we make about our use of fossil fuels and alternative energies, the planet will probably warm by another 1–4 °C on average before the end of the twenty first century (Figure 15.3). The Intergovernmental Panel on Climate Change (IPCC) considers four different *representative concentration pathways* (RCPs), i.e., four different greenhouse gas concentration trajectories expressed in terms of radiative forcing (net energy gain by Earth). An optimistic scenario, assuming declining greenhouse gas emissions for the rest of the twenty first century (RCP2.6, for a 2.6 W/m² radiative forcing), suggests that the global temperature increase could be kept at 1 °C or below. A pessimistic scenario, assuming continuing greenhouse gas emissions (RCP8.5, for an 8.5 W/m² radiative forcing), suggests that Earth could warm by 4 °C or more by the end of the century.

Even though the overall projected temperature increase might seem small in comparison to the wide temperature variations we can experience on a daily or weekly basis (i.e., tens of degrees), it is important to realize that this projection is an average for the entire globe. The temperature increase is expected to be much higher in some places, such as the high latitudes and drier regions of the tropics. It is also thought that a warmer world will feature many more extreme heat events, such as the infamous European heat wave of August 2003 that killed thousands.

Since land masses tend to heat up more than the oceans, which have a greater thermal inertia, we can expect the energy distribution on Earth to change. In fact, temperature changes observed during the twentieth century give us an idea of what to expect in the

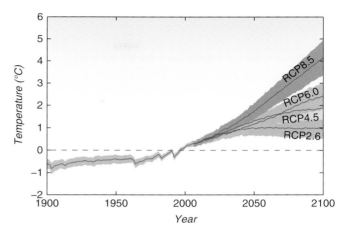

Figure 15.3. Prediction of the future global temperature increase over the next century under four different scenarios. The red line represents the mean over various climate models, while the blue shading represents the standard deviation from the mean. See the text for a description of the representative common pathways (RCPs).

next 100 years. When the temperature was increasing globally by about 1 °C over the last 150 years, it was warming much more in the northern polar region than at lower latitudes (Figure 15.4(b)). By contrast, during the global temperature decrease happening between 0 and 1850 AD, the northern polar region was cooling much more than the rest of the planet (Figure 15.4(a)).

By extrapolation, we can expect the distribution and magnitude of temperature *contrasts* to change even more over the next century, as the planet continues to warm by 1–4 °C. Different temperature contrasts will lead to different transfers of heat to redistribute the energy. Since these transfers of energy are executed in large part by the atmospheric circulation, we can expect global wind systems to be affected, and, as a consequence, daily weather.

15.4 An Altered Water Cycle

We learned in Chapter 5 that the saturation water vapor pressure (e_s) of air increases with temperature, following the Clausius–Clapeyron relationship

(a) 0–1850

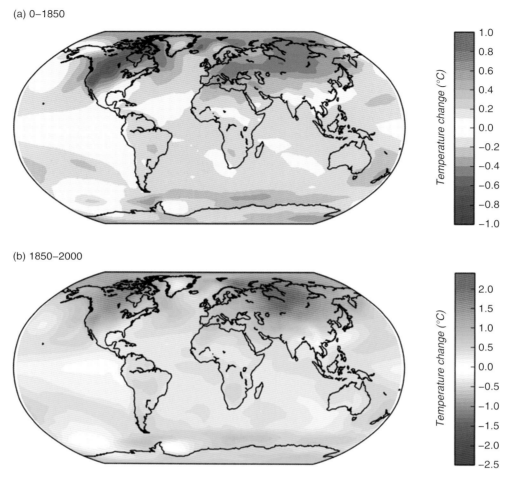

(b) 1850–2000

Figure 15.4. Spatial distribution of temperature changes between (a) 0 and 1850 AD, and (b) 1850 and 2000 AD.

depicted in Figure 5.4. As a rule of thumb, this relationship indicates that a 1 °C increase in temperature results in about a 7% increase in saturation water vapor pressure. Assuming that the global-average relative humidity does not change, which appears to be a good approximation, the evaporative flux from the ocean will increase in a warming world, causing the water vapor content of the lower atmosphere to increase. [Recall that RH = e/e_s. Therefore, for RH to remain constant, e must increase in proportion to e_s. In other words, the actual water vapor content will increase in proportion to the saturation water vapor pressure in a warmer world.] This increase in atmospheric water vapor has several important consequences.

First, recall the water vapor feedback loop (Box 4.5), which tells us that more water vapor means more greenhouse effect, and therefore more warming than from the other greenhouse gases alone. The increase in water vapor will therefore contribute substantially to global warming.

Second, more water vapor in the atmosphere might enhance the formation of clouds. This is subject to high uncertainty in climate models, as cloud formation involves a wide range of scales from aerosols to global wind systems, all of which cannot be resolved in detail. Processes at scales smaller than the climate model grid box (on the scale of tens to hundreds of kilometers) are parameterized in terms of larger-scale features, and these approximations result in uncertainties about the type, number, and composition of clouds. Low clouds, for example, are good reflectors of visible light and good absorbers and emitters of infrared radiation (being lower in the atmosphere, and therefore warmer), whereas high clouds are poor reflectors of visible light and, being higher in the atmosphere, and therefore colder, they emit less radiation than low clouds, while still absorbing infrared radiation effectively. In other words, an increase in high clouds would contribute to global warming (positive feedback), whereas an increase in low clouds would tend to counteract global warming (negative feedback) – although the strength and sign of these cloud feedbacks is still debated in the climate community. In the end, capturing the exact distribution of cloud types is essential in climate modeling, and that is currently an area where climate models differ in details of their forecasts. Moreover, in addition to changes in water vapor, the aerosol content of the atmosphere has also been altered by human activity. In addition to altering the absorption and scattering of radiation by the Earth system (the "direct effect" of aerosols), increasing numbers of small aerosols due to combustion of fossil fuels are changing how water is distributed in clouds. Specifically, the condensed water in clouds is contained in a larger number of smaller droplets, which makes clouds more reflective to solar radiation (the "first indirect effect" of aerosols). Furthermore, changes in the droplet size affect precipitation: more of the smaller cloud droplets and fewer of the larger ones makes collision and coalescence less likely, and in particular less likely to produce precipitating droplets (see Chapter 7). Reduced precipitation is expected to increase cloud lifetime (the "second indirect effect" of aerosols).

Third, more water vapor and more energy in the system might result in changes in precipitation patterns. Climate models suggest a global-average increase of about 2% in precipitation for a 1 °C warming of the surface temperature, which is notably lower than the expected 7% increase for water vapor for such warming (which measures an average amount of water in the atmosphere, whereas precipitation measures the *rate* at which water moves through the atmosphere). The difference appears to involve global energy constraints, where latent heat associated with precipitation must be approximately balanced by atmospheric radiative cooling to space (which is less effective for increasing greenhouse gas concentrations). It appears that this radiative constraint limits the amount of latent heating, and hence condensation,

to precipitation increases of about 2%, rather than 7%. Local changes in precipitation may be much larger, and, in particular, it appears that for extreme precipitation events, such as with hurricanes, precipitation intensity will increase at a rate closer to the 7% increase in water vapor.

Finally, the water cycle will be affected by changes in the general circulation of the atmosphere. Recall from Chapter 9 that the general circulation is driven by the equator-to-pole heat imbalance: low latitudes receive an excess of radiative energy compared to the high latitudes, which drives an overall atmospheric circulation transferring energy from the equatorial and tropical regions to the midlatitudes and polar regions. This takes the form of the Hadley circulation in the tropics and storm tracks in middle latitudes. The temperature contrasts driving the Hadley circulation, and therefore the Hadley cells themselves, are expected to weaken in a warming world. However, even with a weaker circulation, the increase in water vapor is expected to result in more precipitation in the Intertropical Convergence Zone (ITCZ) – which is already relatively rainy. In the subtropics, which are relatively drier, evapotranspiration (i.e., the loss of water from the surface to the atmosphere by evaporation and plant transpiration) associated with warmer temperatures will increase and lead to additional drying.

Thus, overall it appears that precipitation will increase in rainy regions and decrease in dry regions (Figure 15.5). Hence the often used expression, "wet get wetter, while dry get drier." Although somewhat of a simplification, this statement captures the overall direction of the water cycle in a warmer world.

15.5 Changing Global Wind Systems

The Hadley cells will not only become weaker in a warmer world, but also they will expand. This poleward expansion of the Hadley circulation is less well understood, but a robust feature of climate model simulations and in observations during the late twentieth century to the present day. In particular, dry regions associated with the poleward edge of the Hadley circulation in the subtropics appear to be

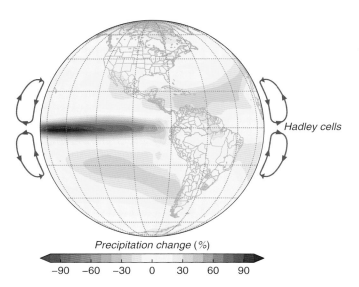

Hadley cells

Precipitation change (%)

-90   -60   -30   0   30   60   90

Figure 15.5. Projected changes in precipitation by the end of the twenty first century.

moving slowly poleward. Additionally, the whole system comprising the two Hadley cells separated by the ITCZ will shift northward. Recall from Chapter 9 that the position of the ITCZ is determined not only by the maximum in insolation, but also by the distribution of land masses, as well as dynamical factors. Since the northern hemisphere contains a larger fraction of land masses (as compared to the larger fraction of ocean in the southern hemisphere), the ITCZ is displaced farther north in summer (e.g., monsoons) than it is displaced southward in (northern) winter. By the same reasoning, the northern hemisphere will warm more in the future, increasing the heat contrast between the two hemispheres, which will displace the ITCZ further north. In other words, the Hadley cells and ITCZ are displaced toward the hemisphere that is warming more, as the atmospheric circulation attempts to redistribute the excess energy from the northern to the southern hemisphere.

Due to similar changes in temperature contrasts, monsoon circulations are expected to intensify, in particular in India and Southern China, with even more precipitation on land.

At higher latitudes and higher altitudes, recall that the strength of the jet stream is determined by the pole-to-equator temperature contrast: as the temperature contrast increases in winter, the jet stream

strengthens; as it decreases in summer, the jet stream weakens. Again, the absolute values of temperature do not matter so much as the temperature *difference* between the equator and the pole. In a warmer world, the Arctic region is expected to warm significantly more than the rest of the planet. Consequently, even though the overall temperature will increase, the pole-to-equator temperature difference will actually *decrease*.

To illustrate, if the temperature at low latitudes is $T_{low} = 28\,°C$ and the temperature at high latitudes is $T_{high} = 4\,°C$, then the temperature difference is

$$\Delta T = T_{low} - T_{high} = 24\,°C.$$

Now, if the low latitudes warm up by $2\,°C$, but the high latitudes warm up by $4\,°C$, then the new temperature contrast is given by

$$\Delta T = 30 - 8 = 22\,°C.$$

The temperature contrast has decreased, and therefore the jet stream will be weaker. What this means for midlatitude cyclones is still unclear at present, in part because cyclones are fueled not only by the pole-to-equator temperature contrast, which is decreasing, but also by latent heat release, which is increasing, as we now describe.

15.6 Midlatitude and Tropical Cyclones in a Warmer World

A weaker jet stream might be associated with weaker midlatitude cyclones; more generally, with a weaker pole-to-equator temperature contrast, with less heat to transfer poleward, we could think that midlatitude cyclones should weaken. However, while extratropical cyclones result largely from the unstable temperature contrast existing in the midlatitudes, they also draw a fair amount of energy from the latent heat released during condensation of water vapor into cloud droplets. As the latent heat is released, it warms the air, warming the column and reducing the surface pressure, causing a stronger storm. One additional complication is that cyclones move heat in the form of water vapor (latent heat), as well as temperature, and, with more water vapor in the atmosphere, storms will have the ability to transport more heat. Therefore, a number of storm processes are affected by global warming, and the impact of the combined effects remains uncertain.

Similarly, in the tropics, tropical cyclones are expected to change in a warming world. We saw in Chapter 12 that the primary energy source for these storms is heat content in the upper ocean, which increases in a warming world. Theoretical studies show that the *maximum potential intensity* of hurricanes, i.e., the maximum amount of energy they can release in optimal conditions, increases with ocean temperature. Thus, we expect that hurricanes forming over warmer water should be able to draw more energy from the ocean by evaporation and to carry more water vapor and more latent heat upward. In turn, latent heat can be released in the atmosphere through condensation, which can fuel more convection and cause more intense hurricanes. However, climate models also project an increase in the *vertical wind shear* in the subtropics, which, as we recall from Chapter 12, inhibits hurricane formation. Thus, two opposing effects will compete in a warmer world, and it is difficult to predict which one will prevail on purely physical arguments. Current expectations are that we might observe fewer hurricanes overall as a result of the increased wind shear, but more of the strongest ones as a result of the higher ocean heat content.

15.7 Beyond Weather

Global warming is predicted to have many other impacts. As warmer oceans expand, for example, and as land glaciers melt, the sea level is rising. Over the past century, the global sea level has risen by about 10–20 cm. About half of that increase is due to the thermal expansion of the oceans. This effect has been amplified locally, such as near the northeast coastline of the US, where weakening ocean currents have led to changes in the slope of the ocean surface.

The two main reservoirs of ice on Earth are Greenland and Antarctica. If the Greenland ice were to melt completely, sea level would rise approximately 6 m, whereas if Antarctica melted completely sea level would rise approximately 60 m. Obviously, such changes in sea level would be catastrophic for human civilization, but neither outcome appears likely for the next 100 years given current climate model simulations. However, polar ice is melting at a faster rate than anticipated by climate model projections from just 5–10 years ago. Although natural variability may explain part of this discrepancy, if it is in fact more directly linked to global warming, then future projections may also underestimate sea level rise. Given that roughly 80% of all humans live within about 100 km of the ocean coastline, changes in sea level are expected to have a dramatic impact on many of the world's major cities. Low-elevation coastal cities, such as Miami in Florida and New Orleans in Louisiana are particularly vulnerable, because they are occasionally affected by storm surge from hurricanes on top of the anticipated rise in sea level.

There is a long list of other climate change impacts. Permafrost (soil that is frozen year round) at high latitude, for example, is thawing. As oceans absorb more CO_2, they are becoming more acidic, which affects their chemistry and biology. Changing temperature and precipitation patterns put plant and animal species under stress. While these changes to the environment do not obviously involve weather directly, they contribute to making climate change a complex issue, and the details of how climate change will proceed under global warming are not yet fully understood.

15.8 The Forecast

In Chapter 13 we discussed how the laws of physics, observations of the atmosphere, and computer programs can be used to predict future weather. We also discussed the limits of that endeavor due to the fundamental uncertainty in observing the atmosphere completely with absolute precision, and how it is important to measure that uncertainty as part of the forecast process. A similar approach is used to make climate model forecasts, but there are some important differences between weather and climate forecast uncertainty that we will discuss here.

Current climate models predict a 1–4 °C temperature increase over the next century (Figure 15.3). Why is there such a large range of uncertainty? One unknown, of course, is our future use of fossil fuels and the corresponding release of carbon dioxide into the atmosphere. Future fossil energy use, however, is not the only source of uncertainty in the forecast. In that regard, the climate forecasting problem shares similarities with, and differences from, the weather forecasting problem. One clear similarity is that both weather forecasting models and climate forecasting models are physically based. They both resolve an initial-value problem and are sensitive to uncertainties in the initial state, as well as uncertainties in the description of key processes in the model. While uncertainty in weather forecasts derives mostly from the initial conditions, in climate forecasts these uncertainties are also linked to many physical processes, but especially to changes in clouds and aerosols, both of which involve processes that are not well represented in the forecasting models and involve the feedback processes discussed earlier. Two key differences between weather and climate forecasting concern the role of the ocean and forcing, both of which affect longer timescales of climate forecasts much more than in weather forecasts.

Climate forecasting requires that we take into account both the atmosphere and the ocean, because they interact on the longer timescales considered in climate modeling, through exchanges of energy, water, momentum, and gases (in particular, greenhouse gases). Therefore, we need a *coupled* model. In contrast, atmospheric motion happens on much shorter timescales than ocean dynamics, and the ocean surface can be described as a static boundary, except in special circumstances, such as near hurricanes.

Including the ocean in climate models is, however, difficult because the ocean – especially the deep ocean – is not nearly as well observed as the atmosphere. Ocean circulations take decades to centuries to redistribute heat, and these slow processes affect the evolution of features in the atmosphere, such as storm tracks and jet streams, on these longer timescales. Thus, the ocean variability simulated by climate models is highly variable, which contributes

to uncertainty in climate forecasts, particularly during the first few decades of the forecast. Uncertainty tends to decrease as one moves from local effects to averages over larger areas. For example, there is often relatively large uncertainty in changes in temperature over local land regions strongly influenced by the ocean, such as West Coast cities, but over larger areas that uncertainty decreases because the local effects become dominated by the main global warming signal.

Forcings are very important for climate models because they can increase predictability on long timescales. Recall from Chapter 13 that, even though we cannot predict weather beyond a couple of weeks due to the chaotic nature of the atmosphere (i.e., sensitivity to the details of the initial state for the forecast), we *can* predict climate on very long timescales because there exist *forcings* that influence the atmosphere on longer timescales than those that dominate weather. In the case of global warming, the forcings from greenhouse gas emissions are known within a range of certainty, and therefore the effects on climate can be predicted.

Among all the factors contributing to climate change, fossil fuel consumption is one that we can control by policy and behavior. If we continue with "business as usual", we are likely heading for the higher end of the 1–4 °C range of temperature increase. If we decide to move away from fossil fuels, however, we might limit the warming to the lower end of the temperature range – although, given the amount of CO_2 already emitted, it is estimated that we will not be able to limit the temperature increase to 2 °C, as was proposed in 2009 at the Copenhagen Conference of the Parties (COP) of the United Nations Framework Convention on Climate Change (UNFCCC).

Summary

Earth's climate has gone through many cold and warm periods in the geologic past. Some of these changes are due in part to cyclical changes in the orbit of Earth around the sun and the tilt of Earth's axis, and to a greenhouse effect feedback due to the release and intake of CO_2 and water vapor initiated by the temperature variations. These effects contribute to the **natural variability** of the climate system.

More recently, Earth's temperature has increased due to **anthropogenic changes** in greenhouse gas concentrations that began with the onset of the Industrial Revolution (**global warming**). The CO_2 concentration, for example, has increased by 40% compared to pre-industrial levels, from about 280 to 400 ppm, while the concentration of methane has increased by 150%.

As a result, the global average temperature has increased by about 1 °C over the past 150 years, and current climate models suggest that the planet will probably warm by another 1–4°C on average before the end of the twenty first century. Moreover, the temperature increase is expected to be much higher in places such as the high latitudes. Extreme heat events are expected to become more frequent, and different temperature contrasts will lead to modifications of weather systems.

The water vapor content of the lower atmosphere will increase in a warmer world, which will lead to:

▶ an enhanced greenhouse effect, leading to more warming than from the increase in CO_2 alone;

▶ changing cloud patterns, although this is subject to high uncertainty in climate models, in particular when it comes to capturing the exact distribution of cloud types and the impact of increasing numbers of small aerosols;

▶ more precipitation: climate models suggest a global-average increase of about 2% in precipitation for a 1 °C warming of the surface temperature, and more for extreme precipitation events, such as with hurricanes; and

▶ weaker Hadley cells and drying in the subtropics, even though precipitation at the ITCZ will

increase due to the increase in water vapor content – "wet get wetter, while dry get drier."

The Hadley cells will also expand, shifting the dry subtropical regions poleward, and the whole system made of the two Hadley cells separated by the ITCZ will shift northward, affecting monsoon circulations, for example.

As the pole-to-equator temperature contrast decreases, the jet stream will be weaker, which may weaken midlatitude cyclones. However, the increase in latent heat release inside storms due to the formation of more cloud droplets (higher water vapor content and higher condensation rate) may cause stronger storms. Thus, the future of midlatitude cyclones remains unclear.

Similarly, a higher ocean heat content will be more favorable for the intensification of hurricanes, but the increase in vertical wind shear will be less favorable for the formation of hurricanes in the first place. Therefore, we might observe fewer hurricanes overall, but more of the strongest ones.

Global warming is predicted to have many additional impacts, from sea level rise and ocean acidification to the melting of Arctic sea ice, permafrost, and the ice sheets in Greenland and Antarctica, among many others.

There are uncertainties in current climate forecasts, linked to physical processes that are not well represented in the forecasting models (e.g., clouds and aerosols), to feedback processes, and to uncertainties in the representation of the ocean, which play a larger role in climate forecasting than in weather forecasting. But it appears that, if we continue consuming fossil fuels at the current rate, we are heading for the higher end of the 1–4 °C temperature range for the global average temperature increase predicted by climate models.

Glossary

absolute humidity a measure of the water vapor content defined as the mass of water vapor per volume of air.

acid rain precipitating water droplets containing sulfuric acid or nitric acid formed by dissolution of sulfur dioxide (SO_2), nitric oxide (NO) and nitrogen dioxide (NO_2), typically released during the combustion of fossil fuels such as oil and coal.

adiabatic cooling temperature decrease experienced by an expanding air parcel as a result of the work exerted by the air parcel molecules on the surrounding air during the expansion process.

adiabatic lapse rate the rate of temperature decrease (increase) experienced by an air parcel ascending (descending) in the atmosphere, as a result of adiabatic expansion (compression). For unsaturated air, the *dry* adiabatic lapse rate is $10\,°C/km$; for saturated air, the *moist* adiabatic lapse rate is about $6\,°C/km$.

adiabatic process a process during which a system experiences internal energy changes due to work, but no transfer of energy or mass takes place between the system and the surrounding environment.

adiabatic warming temperature increase experienced by a descending air parcel as a result of the work exerted on the air parcel by the surrounding air molecules as the air parcel is being compressed.

advection horizontal transport of a quantity (e.g., temperature) by air motion (in the atmosphere) or water flow (in the ocean).

advection fog surface cloud formation by horizontal displacement of relatively warmer and moister air (such as maritime air) over a colder surface (such as cold land or a cold ocean current).

aerosol airborne solid particle of small size, such as contained in smoke, dust, pollen, or pollution.

aggregation process by which two ice crystals collide and stick together to form a larger crystal (as in the formation of snowflakes).

air mass volume of air of large horizontal extent, characterized by relatively uniform temperature and humidity, typically determined by prolonged interaction with the underlying surface (e.g., land or ocean surface at particular latitudes).

air parcel imaginary volume of air with uniform characteristics that does not exchange heat or mass with its surroundings; used especially in comparisons with the surrounding environment when determining the stability of the atmosphere.

albedo percentage of the incident visible light that is reflected by a surface.

Aleutian Low semi-permanent low pressure center located over the Aleutian Islands (between Alaska and Siberia, in the North Pacific Ocean) in average surface pressure maps, especially in winter.

altocumulus a type of cloud forming at mid-elevation in the troposphere (i.e., between 2000 and 6000 m above ground level), and characterized by cumuliform patches separated by clear areas. The patches are typically smaller than in a stratocumulus cloud.

altostratus a type of cloud forming at mid-elevation in the troposphere (i.e., between 2000 and 6000 m above ground level), and taking the form of a relatively uniform gray sheet. The sun can usually be detected through an altostratus as a faint feature.

aneroid barometer a type of barometer in which atmospheric pressure is measured through the volume change of a metallic bellows chamber, which contracts or expands as a result of increasing or decreasing pressure.

angular momentum a quantity defined, in scalar form, as the product of the mass (m), radius of rotation (r), and tangential speed of the rotating body (v): $m \times r \times v$.

anticyclone a high pressure area characterized by weak to no cloud formation, as well as weak to no wind – in particular in the center. Around the anticyclone, winds blow clockwise in the northern hemisphere and counterclockwise in the southern hemisphere.

Archimedes' principle a law of physics indicating that, when submerged in a fluid (partially or totally), a body experiences an upward force equal in magnitude to the weight of the fluid that was displaced by the body. That upward force is also referred to as the *buoyancy force.*

assimilation (data) method by which observations of the atmosphere (made by weather stations and, mostly, satellites) are incorporated (i.e., assimilated) into the mathematical representation of the atmosphere in a numerical weather forecast model, before the model is run for a new forecast cycle.

atmospheric pressure pressure exerted on a horizontal surface by a column of air extending above that surface, due to the effect of gravity on that air column. Atmospheric pressure is typically expressed in hectopascal, millibars, millimeters of mercury, inches of mercury, kilograms per square centimeter, or pounds per square inch.

atmospheric window a frequency band of the electromagnetic spectrum in which the gases constituting the atmosphere absorb little to no radiation (and therefore transmit it).

bar a unit of atmospheric pressure corresponding to 1000 hPa and roughly one atmosphere.

barometer instrument used to measure atmospheric pressure.

Beaufort wind scale a wind scale used in marine meteorology, originally based on observations of the effects of the wind on the sails of a ship, and later extended to include observations of the sea state and land-based effects.

Bergeron process classically called the Wegener–Bergeron–Findeisen process, by which ice crystals grow at the expense of liquid droplets, in the mixed-phase section of clouds, where both ice crystals and supercooled liquid droplets coexist. The process derives from the fact that the saturation vapor pressure over ice is lower than over water, so that air saturated with respect to water is supersaturated with respect to ice.

Bermuda/Azores High a semi-permanent subtropical anticyclone found in the North Atlantic Ocean and centered between the Azores and Bermuda.

black body an idealized body that absorbs all radiation at all wavelengths of the electromagnetic spectrum. It also emits the maximum possible electromagnetic radiation at all wavelengths according to Planck's law.

black ice a thin layer of ice forming when liquid water covering a surface (typically, roads) freezes due to a decrease in temperature, or when liquid precipitation falls upon a surface at subfreezing temperatures and turns to ice. Because the ice is transparent, we can typically see the black asphalt through it, hence its name.

blizzard a snowstorm, characterized by winds greater than 56 km/h (35 mph), which produces blowing and drifting snow that greatly reduces visibility.

buoyancy force experienced by a body submerged, partially or totally, in a fluid, and due to the difference between the pressure exerted by the fluid at the top and the bottom of the body, as described by *Archimedes' principle.*

carbon cycle transfer of carbon through various reservoirs of the Earth system, such as Earth's crust, plants, the ocean, the atmosphere, etc. by various physical processes, such as photosynthesis, respiration, erosion, and volcanic eruptions.

carbon dioxide a variable gas of the atmosphere made of one carbon atom and two oxygen atoms (CO_2) and contributing to the greenhouse effect.

Celsius (degrees) a unit of temperature named after Anders Celsius (1701–1744), who created the first temperature scale based on the freezing point and the boiling point of water as fixed points.

centrifugal force an apparent force experienced by a body in a rotating frame of reference, directed outward from the center of rotation.

chinook wind a relatively warm and dry wind blowing downslope from a mountain or elevated plateau, in which the subsiding air is warmed by adiabatic compression. The chinook is a type of foehn, but the term "chinook" is more often used in the USA and Canada, in reference to the warm and dry winds blowing down from the Rocky Mountains into the Great Plains to the east.

chlorofluorocarbons (CFCs) manufactured chemical compounds widely used as refrigerants, solvents, and in aerosol sprays, and contributing to the breakdown of stratospheric ozone and the so-called ozone hole over Antarctica.

cirrocumulus a type of high cloud taking the shape of a patchy layer, in which the patches are smaller in size than those found in an altocumulus.

cirrostratus A type of high cloud taking the form of a continuous layer through which the sun can be seen, often surrounded by a 22° halo.

cirrus a type of high cloud (found above 5000 m and made of ice crystals) taking the form of thin, wispy strands, sometimes producing virga in the form of falling ice crystals called "mares' tails."

cloud airborne mass of liquid water droplets or ice crystals.

cold front leading edge of a transitional region between a cold and a warm air mass, characterized by a strong temperature gradient and a displacement of the cold air toward the warm air.

cold tongue a region of cold surface water temperatures along the equator extending westward from Peru in the equatorial Pacific Ocean, due to upwelling of deep, cold water.

condensation a phase change from gas to liquid, used in particular when water vapor turns to liquid water when in contact with a surface or a condensation nucleus.

condensation nucleus a small solid particle (of size ~1 μm) that can serve as a "seed" for water vapor molecules to condense and form a water droplet.

conditional instability atmospheric temperature profile for which moist air parcels are unstable only for sufficiently large upward displacement; for smaller displacements, the parcel is stable.

conduction transfer of heat by contact, as kinetic energy is transferred between particles of matter through collision.

conservation of angular momentum (law of) a law of nature according to which the angular momentum of a rotating body remains unchanged, except when acted upon by a torque (a rotational force). When the mass of a rotating object, such as an ice skater, moves toward the axis of rotation, no torque is applied, angular momentum is conserved, and as a consequence the rotation rate increases.

conservation of energy (law of) a law of nature stating that energy cannot be created or destroyed in a system: it can only enter or leave the system, or be transformed into other types of energy inside the system.

conservation of mass (law of) a law of nature stating that mass cannot be created or destroyed in a system: it can only enter or leave the system, or be transformed inside the system.

conservation of momentum (law of) a law of nature stating that the total momentum of a system, where momentum is defined as the product of the mass (or density) and velocity of the bodies constituting the system, remains unchanged unless acted upon by a force, as implied by Newton's laws of motion.

convection vertical displacement of fluid parcels through the fluid due to buoyancy, and, by extension, the vertical transfer of heat, as the internal energy of the parcels is transported with the parcels themselves.

Coriolis force an apparent force experienced by a body in motion in a rotating frame of reference, directed perpendicular to the direction of motion.

cumulonimbus a convective cloud of large vertical development, that typically spans much of the troposphere, flattens out against the tropopause or against a stable layer, and may produce one or more of the following: heavy rain, hail, lightning and thunder, and severe weather, such as tornadoes.

cumulus a low-level convective cloud of relatively small size, forming in a mildly unstable atmosphere, and typically presenting a flat base and a puffy, cauliflower-like appearance.

cumulus congestus a convective cloud of significant vertical development, forming in a moist, unstable atmosphere, and typically merging with surrounding convective clouds to obstruct the horizon.

cyclone a circulation developing around a low pressure region of the atmosphere, and taking the form of a counterclockwise (in the northern hemisphere) or clockwise (in the southern hemisphere) vortex.

density a measure of the concentration of a substance defined as its mass per unit volume.

deposition a phase change by which matter changes state from gas to solid, used in particular when water vapor turns to ice.

deterministic forecast a type of weather forecast based on the computational solution of prognostic equations describing the future evolution of the atmosphere based on physical principles.

dew formation of water droplets by condensation of water vapor onto solid surfaces, typically onto low-lying elements such as grass.

dew point depression difference between the actual (measured) temperature and the dew point temperature.

dew point temperature temperature at which an air parcel becomes saturated if the air is cooled at constant pressure.

diurnal cycle a cyclical pattern of 24 hour period. Although it applies to many quantities, it is typically used to describe the variation in temperature due to the daily cycle in incoming solar radiation.

Doppler radar a type of radar using the frequency shift measured in the return signal to infer the speed of motion of precipitating particles relative to the radar.

drizzle a type of precipitation characterized by water droplets smaller than 0.5 mm in diameter.

dry line boundary between two air masses of different moisture content as defined by the dew point.

dynamic equilibrium a form of equilibrium in which the contents of the system might be changing in time, but not in overall amount, because the inputs are balanced by the outputs and there is no net change in the system.

El Niño/Southern Oscillation (ENSO) a semi-periodical variation in the coupled oceanic and atmospheric circulation taking place in the equatorial region of the Pacific Ocean.

electromagnetic wave oscillation of the electric and magnetic fields that propagates through space at the speed of light, characterized by its frequency, or wavelength, and corresponding to different types of radiation.

emission spectrum intensity of electromagnetic radiation emitted by a body at each frequency, or wavelength.

ensemble forecast a type of forecast in which more than one computer simulation is made in order to estimate the uncertainty in the forecast. Each simulation is called an ensemble member, and consists of small changes to the initial conditions and/or the computer model.

entrainment mixing of dry, unsaturated, environmental air into the saturated regions of a cloud, resulting in the evaporation of cloud droplets.

environmental lapse rate rate of decrease of temperature with altitude, measured in the atmosphere at a given time and location.

evaporation a phase change by which matter changes state from liquid to gas, used in particular when liquid water turns to water vapor at a water–air interface, such as Earth's surface (ocean, lake, river, etc.) or the surface of a water droplet (cloud droplet, raindrop).

evaporation fog a type of fog occurring when water evaporates from a warm water surface into colder air and brings it to saturation.

extratropical cyclone a midlatitude weather system in which winds circulate air around a low pressure center and transport energy from lower to higher latitudes through the equatorward movement of cold air and poleward movement of warm air associated with cold and warm fronts, respectively.

extratropics latitudinal band found poleward of the subtropics, also referred to as the midlatitudes.

eye central region of a tropical cyclone, surrounded by the eyewall and characterized by subsidence, relatively little middle and high cloud, and relatively lighter winds.

eyewall ring of convective clouds surrounding the eye of a tropical cyclone.

Fahrenheit (degrees) a unit of temperature named after Gabriel Fahrenheit (1686–1736), who designed the first mercury thermometer and invented a temperature scale based on three fixed points: 0 in a mixture of ice, water and sea salt; 32 in water and ice; and 96 "in the mouth or armpit of a healthy man."

Ferrel cell a feature of the general circulation of the atmosphere describing the meridional overturning circulation in the midlatitudes, between the Hadley cell and the polar cell.

foehn a warm and dry wind resulting from the flow of air over a mountain and down the leeward slope, as the air subsides and warms at the dry adiabatic lapse rate.

fog formation of cloud droplets in the lower atmosphere, near the surface, due to a temperature decrease (radiation fog and advection fog) or an increase in water vapor content to saturation (evaporation fog).

fragmentation process by which large ice crystals break into many small parts that can then serve as ice nuclei to seed new ice crystals.

freezing nucleus a particle that can initiate the growth of an ice crystal from supercooled liquid water droplets.

freezing rain rain that falls as liquid water raindrops, but freezes upon impact with a cold surface at a temperature at or below 0°C, and forms a thin layer of ice (typically "black ice" on roads).

friction *See surface friction.*

front a transitional region between a cold and a warm air mass, characterized by a strong temperature gradient, a pressure trough, and a wind shift.

funnel cloud a rotating cloud emerging from the wall cloud at the base of a cumulonimbus and extending downward toward the ground in the process of tornado formation.

Fujita scale a tornado intensity scale designed by Tetsuya Fujita (1920–1998) that ranks the intensity of tornadoes using wind speed and the damage caused by tornadoes.

general circulation large-scale, time-average organization of air currents around Earth.

geostationary satellite a satellite in orbit around Earth, in the equatorial plane and at such a distance (35,786 km) that the satellite orbits at the same angular velocity as Earth, so it remains positioned above the same location on Earth.

geostrophic balance state of equilibrium in which the air moves in a straight line along the isobars (height contours) due to a force balance between the pressure gradient force and the Coriolis force, which are equal and opposite, and directed at right angles to the direction of motion.

geostrophic wind conceptual wind speed and direction assuming the pressure gradient and Coriolis forces balance, whereby the wind flows parallel to isobars with a speed proportional to the magnitude of the pressure gradient. Useful for quickly estimating the actual wind from isobars and height contours.

global warming enhancement of the greenhouse effect due to the anthropogenic increase in greenhouse gases such as carbon dioxide and methane.

GOES US Geostationary Operational Environmental Satellite system that has provided satellite imagery to support weather forecasting and research since the 1970s.

gradient wind balance state of equilibrium in which the air moves in a straight line along the isobars (height contours) due to a force balance between the pressure gradient force, the Coriolis force, and the centrifugal force due to the curvature of the flow, which are all directed at right angles to the direction of motion.

graupel a type of precipitation in which typically conical, "spongy" ice pellets form by riming of supercooled liquid water droplets freezing upon impact as the pellets fall through the cloud.

greenhouse effect radiative effect by which Earth's surface is at a higher temperature than it would be in the absence of greenhouse gases (water vapor, carbon dioxide, methane, ozone, etc.), due to the absorption and emission of infrared radiation by these gases.

gust front leading edge of the outflow of cold, dense air spreading downward and outward from a thunderstorm.

Hadley cells a feature of the general circulation of the atmosphere describing the average circulation of air in the subtropics, as warm, moist air rises near the equator, flows poleward near the tropopause, subsides in the subtropics, and returns to the equator near the surface.

hail a type of frozen precipitation formed by the freezing of successive layers of ice as the ice pellet travels repeatedly through the updraft of a cumulonimubus cloud collecting water droplets, cloud droplets, snow, and ice crystals in layers with each pass through the updraft.

halo optical phenomenon appearing as a ring of light around the sun due to the refraction and reflection of light by ice crystals, such as found in a cirrostratus. Halos form at 22° and 46° angles from the sun, and the 22° halo is most common.

haze suspended, dry, solid particles, such as dust, smoke, and pollution, obscuring the sky and reducing visibility.

heat amount of energy that moves between bodies due to temperature contrasts. On average, heat transfers energy from warm to cold objects.

heat capacity amount of energy required to raise the temperature of a mass of matter by one degree Celsius.

hectopascal a unit of pressure, used in particular to express atmospheric pressure. For example, the average sea level pressure on Earth is 1013 hPa.

heterogeneous nucleation formation of a liquid water droplet by condensation of water vapor onto a surface, such as provided by a condensation nucleus (i.e., an aerosol).

homogeneous nucleation formation of a liquid water droplet by condensation of water vapor molecules onto themselves, without seeding by an aerosol or a collecting surface.

hurricane *See tropical cyclone.*

hygrometer instrument used to measure the water vapor content of the surrounding air.

ice nucleus a small particle that can serve as a seed for an ice crystal to grow.

Icelandic low a semi-permanent low pressure region found near Iceland and most pronounced in winter.

ideal gas law equation of state describing the relationship between the pressure, volume, mass (or density), and temperature of an "ideal" gas (where molecular collisions are perfectly elastic), typically expressed as $pV = nR^*T$, or $p = \rho RT$, where p is the pressure, V is the volume, n is the number of molecules in volume V, T is the temperature, ρ is the density, R^* is a universal constant that applies to any gas, and R is a gas constant specific to the atmosphere.

Indian monsoon a tropical atmospheric circulation resulting from the displacement of the ITCZ and Hadley cells in the Indian Ocean, and enhanced by local geographical and dynamical features. Importantly, the Indian monsoon alternates between a wet season (when the winds are onshore) and a dry season (when the winds are offshore).

infrared a frequency band of the electromagnetic radiation spectrum corresponding to wavelengths in the 1–1000 μm range (shorter wavelengths than microwaves, but longer than ultraviolet).

infrared window a frequency band in the infrared region of the electromagnetic radiation spectrum in which atmospheric gases do *not* significantly absorb.

initial conditions initial state of a numerical weather forecast model before it is integrated forward in time, in which the most recent observations have been assimilated.

Intertropical Convergence Zone (ITCZ) a region of the atmosphere found near the equator, characterized by convergence of the trade winds and convection.

inversion (temperature) increase in temperature with height, appearing as a positive slope in a vertical temperature profile, and therefore as a deviation from the normal decrease of temperature with height in the troposphere. By extension, the term also refers to the layer in which the temperature increase with height is occurring.

ionosphere region of the atmosphere where atoms and molecules are ionized (i.e., lose electrons) by solar radiation, thus affecting the propagation of radio waves.

isobar a line of constant pressure.

isotherm a line of constant temperature.

jet stream upper-level westerly air current found near the tropopause, between 30 and 60 degrees of latitude, in the northern and southern hemispheres. The jet stream typically has meanders and locally maximum wind speeds, and sometimes splits into two jets: a polar jet stream and a subtropical jet stream.

katabatic winds a dense, negatively buoyant air current flowing downslope from a mountain or an elevated plateau under the effect of gravity. The term is typically used to describe fast, cold, dense air currents flowing down from the elevated ice sheets of Antarctica and Greenland.

kelvin a unit of temperature named after Lord Kelvin (1824–1907). The Kelvin scale starts at absolute zero ($-273.16\,°C$) and increases with intervals of the same magnitude as the Celsius scale (i.e., $1\,K = 1\,°C$).

kinetic energy energy associated with motion that forms the basis for defining temperature as a measure of the average kinetic energy of a large number of air molecules.

knot a unit of wind speed, defined as one nautical mile per hour: $1\,knot = 1.852\,km/h = 1.151\,mph$.

land breeze a wind blowing from land to sea when the water is warmer than the adjacent land, due to an overturning circulation caused by the temperature contrast.

lapse rate rate of temperature decrease with height.

La Niña "cold" phase of El Niño/Southern Oscillation corresponding to the maximum extent of the cold tongue in the equatorial Pacific Ocean.

latent heat amount of energy absorbed or released when matter changes phase between the solid, liquid, and gas states while maintaining constant temperature.

lenticular cloud elongated cloud, shaped like a lens, forming over the top of a mountain as the horizontal flow of moist air is deflected upward by the summit, causing adiabatic cooling and bringing the air to saturation.

level of free convection (LFC) altitude at which a lifted air parcel becomes warmer than its environment and rises by the sole action of the buoyancy force.

lifting condensation level (LCL) altitude at which a rising air parcel reaches saturation due to adiabatic expansion and cooling.

lightning a discharge of electricity resulting from the separation of electrical charge occurring when hailstones collide with smaller ice particles in a cumulonimbus cloud.

mammatus cloud a type of cloud forming on the underside of a cumulonimbus anvil, as dense air, cooled by evaporation, entrains cloud particles from the anvil downward. A rare example of saturated descent of air.

mean sea level average level of the sea surface, used as a reference to measure, for example, atmospheric pressure, so that all measurements are comparable.

medium-range forecast a weather forecast established for the 3–10-day time range, as opposed to short-range forecasting for the next 3 days.

mercury barometer instrument used to measure atmospheric pressure, made of an inverted glass tube containing mercury up to the level where the weight of the column of mercury is balanced by the atmospheric pressure applied to the surface of the mercury in the reservoir.

mesopause transition region between the mesosphere and the thermosphere.

mesosphere layer of the atmosphere from about 50–90 km (1 hPa to 0.01 hPa), in which the ozone concentration decreases with height, along with temperature, up to the mesopause.

meteogram a graph showing the time evolution of variables measured at a weather station (temperature, dew point, atmospheric pressure, relative humidity, wind, etc.).

methane a variable gas of the atmosphere made of one carbon atom and four hydrogen atoms (CH_4) and contributing to the greenhouse effect.

microburst isolated, short-lived, but very intense downdraft occurring under a severe thunderstorm. These downdrafts can reach up to 150 knots, and pose a serious hazard for aircraft during take-off and landing.

midlatitudes The 30°–60° latitude region found between the Hadley cell and the polar cell in each hemisphere.

midlatitude cyclone *see extratropical cyclone.*

millibar a unit of atmospheric pressure equal to one hectopascal. For example, the average sea level pressure on Earth is 1013 mb = 1013 hPa.

mixing ratio a measure of the water vapor content of the air, defined as the mass of water vapor per mass of dry air.

momentum a quantity defined as the product of the mass and velocity of a body in motion.

monsoon a seasonal, thermally driven atmospheric circulation due to uneven heating of continents and ocean, associated with changes in surface wind direction, precipitation, and temperature.

neutral stability a state of the atmosphere in which displaced air parcels neither rise nor sink, but remain at the same height.

nimbostratus a rain-producing stratus cloud.

nitric oxide gas, rapidly oxidized to nitrogen dioxide (NO_2) in the atmosphere and contributing to the formation of acid rain; nitric oxide (NO) and nitrogen dioxide (NO_2) are jointly referred to as NO_x.

nitrogen a permanent gas made of two nitrogen atoms (N_2) and constituting about 78% of the atmosphere.

nitrogen dioxide gas, oxidized to nitric acid (HNO_3) in the atmosphere and contributing to the formation of acid rain; nitric oxide (NO) and nitrogen dioxide (NO_2) are jointly referred to as NO_x.

nitrous oxide a variable gas of the atmosphere made of two nitrogen atoms and one oxygen atom (N_2O) and contributing to the greenhouse effect.

occluded front boundary between two relatively cold air masses, resulting from the closing of the warm sector in a midlatitude cyclone and lifting of the warm air mass aloft. Appears on a surface map as a trough in the pressure field and an axis of warm air extending from the triple point to the low center.

orographic precipitation precipitation resulting from the ascent of air along the slopes of a mountain or mountain range.

overrunning ascent of warm air over a relatively cooler air mass along the warm front of a midlatitude cyclone.

oxygen a permanent gas of the atmosphere made of two oxygen atoms (O_2) and constituting about 21% of the atmosphere.

ozone a variable gas of the atmosphere made of three oxygen atoms (O_3). Tropospheric ozone is detrimental to human health and can cause severe respiratory problems. Stratospheric ozone, however, absorbs ultraviolet radiation, and thus protects us from the damaging effects of ultraviolet energy (in particular, skin cancer).

ozone hole region of the stratosphere, above Antarctica, where the destruction of the ozone layer by chlorine compounds over the last 50 years has been most pronounced.

ozone layer region of the stratosphere where the concentration of ozone increases and reaches a maximum, at an altitude of about 50 km.

Pacific High a semi-permanent high pressure region found over the North Pacific Ocean near Hawaii.

Pacific warm pool western region of the equatorial Pacific Ocean, characterized by higher temperatures than found in the eastern basin (i.e., in the cold tongue).

parameterization mathematical characterization of small-scale processes, which cannot be fully resolved in a numerical model, in terms of larger-scale processes and variables.

permanent gases gases whose relative concentration (relative to each other) does not vary much over time and space, up to about 80 km. For example, oxygen always represents about 21% of atmospheric gases.

phase a state of matter: solid, liquid, or gas.

photosynthesis chemical transformation of water and carbon dioxide into organic matter, given the appropriate amount of radiative energy (sunlight).

polar cell overturning circulation occurring between about 60° of latitude and the pole in each hemisphere, as part of the general circulation of the atmosphere.

polar front pronounced meridional temperature gradient found in the midlatitudes in each hemisphere.

polar-orbiting satellite a satellite orbiting Earth from pole to pole at about 800 km altitude.

ppm parts per million, a unit used to express gas concentration.

precipitation liquid or solid water particles falling faster than the updrafts in which they are embedded.

pressure a quantity characterizing the frequency and intensity of average molecular collision, defined as the net force per unit area.

pressure gradient force force resulting from the existence of a pressure gradient that, in the absence of other forces, acts to accelerate air from high to low pressure.

prognostic equations equations from the laws of physics describing the evolution of a set of variables such as temperature, humidity, and wind, which allow us to calculate (i.e., make a *prognosis* of) the future state of the atmosphere.

radar instrument that estimates the location and size of precipitating solid or liquid water particles based on the timing and intensity of returning electromagnetic waves after they have been emitted by the radar and reflected by the particles.

radiation oscillation of the electromagnetic field that manifests itself as waves of various wavelengths propagating at constant speed (i.e., the speed of light).

radiation fog formation of cloud droplets due to the decrease in temperature resulting from a loss of energy by emission of infrared radiation near the ground (synonymous with "ground fog").

radiative equilibrium a state whereby a body emits the same amount of radiative energy as the amount it absorbs – although not necessarily at the same wavelength.

radiosonde instrument that measures variables of the atmosphere as it ascends with a helium-filled balloon, and radios the measurements back to a receiving station.

rain precipitating liquid water particles.

rain gauge instrument that measures the amount and, for certain designs, the rate of liquid precipitation.

reflection process by which the direction of propagation of electromagnetic waves reverses at the incident angle, without absorption, when the waves encounter an obstacle.

relative humidity a measure of the water vapor content defined as the percentage of water vapor present in the air, compared to the maximum amount at saturation (i.e., actual water vapor pressure divided by saturation water vapor pressure, e/e_s).

respiration a chemical reaction by which organic matter is broken down into water and carbon dioxide in the presence of oxygen, and energy is released.

ridge (pressure) axis of relatively high values in a pressure field without reaching a local maximum, analogous to a mountain ridge on a topographic map.

riming a process by which supercooled liquid water droplets freeze onto an existing ice crystal during a collision – a process also called accretion, which tends to form conical pellets called graupel.

Saffir–Simpson hurricane wind scale a scale used in tropical meteorology to indicate the intensity of hurricanes (i.e., tropical cyclones) based on their maximum sustained wind speed.

saturation a state of equilibrium in which the amount of water vapor occupying a given space is maximum.

saturation water vapor pressure partial pressure of water vapor at saturation; it is proportional to temperature according to the Clausius-Clapeyron equation.

scattering reflection of electromagnetic radiation in multiple directions.

sea breeze a wind blowing from sea to land when the land is warmer than the adjacent sea, due to an overturning circulation caused by the temperature contrast.

seasonal cycle cyclical variation of period one year of the atmospheric variables due to the changes Earth experiences in its exposure to solar radiation as it revolves around the sun.

selective absorber a molecule that absorbs radiation at specific wavelengths, but not at others – in contrast to a black body, which absorbs all incident radiation.

shelf cloud a cloud forming over the leading edge of an advancing gust front, as warmer air is lifted by the cool air spreading outward underneath a thunderstorm.

sleet (ice pellets) precipitating frozen water drops, which are in liquid form at some point during their fall, but freeze before touching the ground as they fall through a colder layer.

smog formation of cloud droplets (fog) in a particularly polluted environment, i.e., in the presence of numerous aerosols and chemicals, such as those produced by industries (historically, smoke in London).

snow precipitating ice crystals and ice-crystal aggregates that formed in the cold part of a cloud and did not melt before reaching the surface.

specific humidity a measure of humidity defined as the mass of water vapor per mass of moist air.

squall line a line of thunderstorms usually associated with an advancing front.

stable (air, atmosphere) a state of the atmosphere in which displaced air parcels return to their original altitude when released.

station model a set of conventional symbols used to synthesize the measurements and observations made at a weather station or buoy.

stationary front transitional region between a cold and a warm air mass when neither mass is advancing against the other.

storm tracks preferential locations for midlatitude cyclone development, in particular over the Western Pacific and West Atlantic oceans, due to a combination of factors (a strong meridional temperature gradient, prominent land masses and land–ocean temperature contrasts, and a weakly stable atmosphere).

stratocumulus a type of low cloud characterized by an overall stratiform appearance, but broken up in patches, typically due to the presence of mild convection or turbulence.

stratopause boundary between the stratosphere and the mesosphere.

stratosphere stable layer of the atmosphere found between about 10 and 50 km where the temperature increases due to the presence of ozone and the absorption of ultraviolet radiation.

stratus a type of cloud forming in the lower troposphere (i.e., below 2000 m altitude), and taking the form of a uniform gray sheet.

subgeostrophic wind wind occurring when the flow is curved and around a low pressure region, or trough, and so that the wind speed is less than the geostrophic wind due to flow curvature.

subpolar low a low pressure region forming on the poleward side of the midlatitudes, typically between 50° and 70° of latitude, as in the Aleutian Low and the Icelandic Low observed in average pressure fields.

sublimation phase change by which matter changes state from solid to gas, without going through the liquid state (in meteorology, from ice to water vapor).

subsidence downward motion of air, accompanied by adiabatic compression and warming.

subtropical high a semi-permanent high pressure region found between 20° and 40° of latitude and over the ocean, and strongest in summer.

subtropics latitudinal belt found in each hemisphere on the poleward side of the tropics, and considered to be between about 20° and 40° of latitude, with some geographical variations in places.

supercell thunderstorm a long-lived rotating thunderstorm with updrafts separated from downdrafts. Hail often occurs with these storms, and occasionally tornadoes.

supercooled (water) water cooled below 0 °C without a change of state from liquid to solid (i.e., without freezing).

supergeostrophic wind wind occurring when the flow is curved and around a high pressure region, or ridge, and so that the wind speed is greater than the geostrophic wind due to flow curvature.

supersaturation a state in which the air contains more water vapor than the maximum capacity allows at that temperature, i.e., when the water vapor pressure is greater than the saturation water vapor pressure ($e > e_s$), which means that the relative humidity is greater than 100%.

surface friction drag on air motion due to roughness elements at the surface, which tends to reduce the speed of the wind close to the surface, among other effects.

sustained wind speed of the wind averaged over a standard time period, such as one minute.

synoptic scale a spatial scale encompassing weather events on the order of a few hundreds to a few thousands of kilometers (e.g., typical high and low pressure systems in the midlatitudes).

teleconnection remote influence of weather events on distant locations in the atmosphere.

temperature a measure of the heat content of air, therefore related to the kinetic energy of air molecules.

thermal cloud-free ascending motion occurring in a convective cell, typically on sunny afternoons over open terrain.

thermocline region of strong temperature gradient separating the well-mixed and relatively warmer surface layer of the ocean from the deeper and colder ocean.

thermometer instrument used to measure temperature.

thermosphere upper layer of the atmosphere, above about 100 km, in which air density is extremely low and the motion of individual air molecules extremely fast.

thunder a sound shock wave resulting from the rapid heating and expansion of the channel of air through which electricity flows during a lightning strike.

thunderstorm a cumulonimbus cloud characterized by strong updrafts and downdrafts, and usually producing heavy showers, lightning, thunder, and possibly hail.

tornado a rapidly rotating vortex occurring with cumulonimbus clouds, tens to hundreds of meters in diameter, characterized by extremely low central pressure and high winds of up to 250 knots and sometimes more.

Tornado Alley a region of the US Midwest where environmental conditions and atmospheric dynamics are favorable for the formation of tornadoes.

trade winds surface winds blowing at low latitude as part of the lower branch of the Hadley cells. On average, trade winds are northeasterly in the northern hemisphere, southeasterly in the southern hemisphere, and converge at the ITCZ.

tropical cyclone tropical weather system developing over the ocean, deriving its energy from the evaporation of ocean water, and taking the form of a vortex of strong winds (sustained winds greater than 64 knots) blowing cyclonically around a low pressure center (the eye), with spiraling bands of cumulonimbus clouds producing heavy rain.

tropical depression a precursor to a tropical storm in the Saffir–Simpson hurricane intensity scale, with 20–34-knot winds.

tropical disturbance a precursor to a tropical depression in the Saffir–Simpson hurricane intensity scale, with 0–20 knot winds.

tropical easterly wave a trough of low pressure in the tropical easterly flow in which thunderstorms congregate on the east side of the trough axis due to convergence and enhanced convection. Under favorable circumstances, a tropical easterly wave can evolve into a tropical storm.

tropical storm a precursor to a tropical cyclone in the Saffir-Simpson hurricane intensity scale, with 34–64-knot winds.

tropics latitudinal belt delimited, broadly, by the Tropic of Cancer in the northern hemisphere and the Tropic of Capricorn in the southern hemisphere.

tropopause boundary between the troposphere and the stratosphere.

troposphere first layer of the atmosphere extending from the surface to the altitude where temperature stops decreasing with height (the tropopause). Essentially all weather occurs in this layer.

trough (pressure) axis of relatively low values in a pressure field without reaching a local minimum, analogous to a valley on a topographic map.

typhoon name given to a tropical cyclone when it forms over the western North Pacific Ocean.

ultraviolet a frequency band of the radiation spectrum corresponding to wavelengths in the 0.005–0.4 μm (5–400 nm) range (shorter wavelengths than visible radiation, but longer than x-rays).

uncertainty range of possible values (error) when a variable is measured (due to instrumental errors) or predicted (due to errors in the initial conditions and model errors).

unstable (air, atmosphere) a state of the atmosphere in which air parcels that are displaced accelerate in the direction of the displacement due to a buoyancy force. For upward displacement, this means an upward buoyancy force and upward acceleration.

upwelling vertical displacement of water from the deep ocean to the surface due to the divergence of surface waters.

valley breeze warming and upward displacement of air along the slopes of a valley during the day, when the slopes are warmed by the absorption of sunlight.

valley fog formation and concentration of fog at the bottom of a valley, due to the downslope drainage of cold, dense air into the valley at night, when the slopes cool by emission of infrared radiation.

variable gases gases whose concentration varies in the atmosphere over space and time, such as water vapor and carbon dioxide.

virga precipitating particles that evaporate or sublimate before reaching the surface.

Walker circulation a large-scale circulation of the atmosphere above the equatorial Pacific Ocean due to the temperature contrast between the Pacific warm pool and the equatorial cold tongue.

wall cloud a rotating cloud at the base of a cumulonimbus (i.e., a thunderstorm) preceding the possible formation of a funnel cloud and tornado.

warm front warm edge of a transitional region between a warm and a cold air mass, characterized by a strong temperature gradient, a pressure trough, wind shift, and the overrunning of warm air over cold air producing clouds and precipitation.

warm sector region delimited by the cold and warm fronts in a midlatitude cyclone.

water vapor water in the gas phase.

water vapor pressure a measure of the amount of water vapor in the atmosphere defined as the pressure exerted by water vapor alone, as if all other gases were absent.

wavelength distance between two equal phases of a wave, such as two crests or two troughs.

weather station assemblage of various weather instruments at conventional heights or, more generally, the location where these meteorological observations are made.

wind gust a local, sudden, and brief increase in wind speed due to the transport of higher momentum air from a different altitude by turbulent eddies.

wind shear a change in the wind in a given direction, often referring to a change in wind speed in a direction normal to the direction of the wind, e.g., a vertical increase in horizontal wind, or a meridional increase in zonal wind.

wind vane instrument used to measure wind direction.

References

Dee, D. P., Uppala, S. M., Simmons, A. J. *et al.* (2011). The ERA-Interim reanalysis: configuration and performance of the data assimilation system. *Quart. J. Roy. Meteorol. Soc.*, **137**, 553–597, doi:10.1002/qj.828.

Frank, D. C., Esper, J., Raible, C. C. *et al.* (2010). Ensemble reconstruction constraints on the global carbon cycle sensitivity to climate. *Nature*, **463**, 527–532, doi:10.1038/nature08769.

Hakim, G. J., Emile-Geay, J., Steig, E. J. *et al.* (2016). The Last Millennium Climate Reanalysis Project: framework and first results. *J. Geophys. Res. Atmos.*, **121**, doi:10.1002/2016JD024751.

Jouzel, J., Masson-Delmotte, V., Cattani, O. *et al.* (2007). Orbital and millennial Antarctic climate variability over the past 800,000 years. *Science*, **317**(5839), 793–797, doi:10.1126/science.1141038.

Lemon, L.R. and Doswell, C.A. (1979). Severe thunderstorm evolution and mesocyclone structure as related to tornadogenesis. *Month. Weather Rev.*, **107**, 1184–1197.

Lord, S. D. (1992). A new software tool for computing Earth's atmospheric transmission of near- and far-infrared radiation. NASA Technical memorandum 103957.

Credits

Satellite images, land weather station, and buoy measurements were obtained from the National Climate Data Center (www.ncdc.noaa.gov) at the National Oceanic and Atmospheric Administration (NOAA). Radiosonde measurements were obtained from the Earth System Research Laboratory (esrl.noaa.gov) at NOAA. ERA-Interim numerical analyses were obtained from the European Centre for Medium-Range Weather Forecasts (ECMWF) after Dee *et al.*, The ERA-Interim reanalysis: configuration and performance of the data assimilation system, Quarterly Journal of the Royal Meteorological Society (2011). Infrared transmission spectra were obtained from the Gemini Observatory website (www.gemini.edu) after Lord (1992). NASA images were obtained from the Earth Observatory (earthobservatory.nasa.gov).

All maps were created by the authors using Python version 3.4 (www.python.org). All photographs have been provided by the authors, except for those in the list below.

4.1.3	University of Frankfurt / Wikimedia Commons / Public Domain
4.1.4	Image from Nobel Media AB 2014 / Wikimedia Commons / Public Domain
4.1.5	Image from Wikimedia Commons / Public Domain
4.2.1	Image from NASA / Public Domain
4.2.2	Image from Caroline Planque
4.6	Data from the Gemini Observatory after Lord (1992)

Chapter 5

Title banner	Image from Caroline Planque
5.5	Image from Freerange Stock / Public Domain
5.6	Image from Aleksandar Momirovic / Freerange Stock / Public Domain
5.7	Data from ECMWF
5.8	Data from NOAA
5.9	Data from NOAA

Chapter 6

Title banner	Image from Tim Mossholder / Pexels / Public Domain
6.1.1	Getty Images
6.1.3	Image from Caroline Planque
6.9(a)	Image from NASA
6.10	Image from NASA
6.21	Image from Caroline Planque
6.23	Getty Images
6.26	Image from NASA
6.29	Image from Wikimedia Commons / Public Domain
6.30	Image from Caroline Planque
6.31	Image from Caroline Planque
6.35	Getty Images

Chapter 7

Title banner	Image from Caroline Planque
7.6	Image from Caroline Planque
7.11	Image from Reid Wolcott
7.15	Image from Caroline Planque

Chapter 8

8.15	Data from NOAA
8.16	Data from ECMWF
8.20	Image from NASA

Chapter 9

Title banner	Courtesy of the Museum of the History of Science, Oxford, UK
9.1	Data from NOAA
9.2	Data from NOAA
9.3	Data from NOAA
9.4	Data from NOAA
9.5	Data from NOAA
9.6	Data from NOAA
9.7(b)	Data from NOAA
9.9	Image from NASA
9.20	Data from NOAA

Chapter 10

Title banner	Image from NASA
10.4(a)	Image from NASA
10.5(a)	Image from NASA
10.13	Data from ECMWF and NASA data
10.14	Data from ECMWF and NASA data
10.15	Data from ECMWF and NASA data
10.16	Data from ECMWF and NASA data
10.19(a)	Data from ECMWF
10.19(b)	Image from NASA
10.20	Image from the American Meteorological Society
10.21	Image from the American Meteorological Society
10.22	Image from the American Meteorological Society
10.23	Image from the American Meteorological Society

Chapter 11

Title banner	Image from Caroline Planque
11.3	Getty Images
11.4	Image from Uli Feuermeister / Wikimedia Commons / Public Domain
11.6	Image from skeeze / Pixabay / Public Domain
11.7	Image from Sensenmann / Wikimedia Commons / Public Domain

11.9	Image from Jared Davidson / Freerange Stock / Public Domain
11.13	Image adapted from Lemon and Doswell (1979)
11.14	Getty Images
11.15	Getty Images
11.16	Getty Images
11.17	Getty Images
11.18	Getty Images

Chapter 12

Title banner	Image from NASA
12.1(a)	Image from NASA
12.2	Data from NOAA and Remote Sensing Systems
12.3	Image from NASA
12.6	Data from ECMWF
12.7(a)	Data from ECMWF
12.7(b)	Data from ECMWF and NASA data
12.8	Data from NOAA
12.10	Data from ECMWF and EUMETSAT data

Chapter 13

Title banner	Courtesy of the Department of Atmospheric Sciences at the University of Washington
13.3	Data from NOAA
13.4	Data from NOAA
13.5	Data from NOAA
13.6	Images from NASA
13.7(a)	Data from ECMWF
13.7(b)	Image from NASA

13.7(c)	Data from ECMWF
13.8	Image from NOAA

Chapter 14

Title banner	Image from Foto-RaBe / Pixabay / Public Domain
14.2	Image from NASA
14.3(b)	Image from czu_czu_PL / Pixabay / Public Domain
14.3(d)	Image from David Karich / Pixabay / Public Domain
14.3(f)	Image from Igors Jefimovs / Wikimedia Commons / Public Domain
14.4	Getty Images

Chapter 15

Title banner	Image from Caroline Planque
15.1	Data from NOAA, after Jouzel *et al.* (2007)
15.2(a)	Data from Hakim *et al.* (2016), Last Millennium Climate Reanalysis Project
15.2(b)	Data from NOAA, after Frank *et al.* (2010)
15.3	Data from the IPCC, courtesy of the Institute for Atmospheric and Climate Science, ETH Zürich
15.4	Data from Hakim *et al.* (2016), Last Millennium Climate Reanalysis Project
15.5	Data from the IPCC, courtesy of the Institute for Atmospheric and Climate Science, ETH Zürich

Index